200 Challenging Math Problems

every 6ᵗʰ grader should know

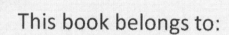

This book belongs to:

....

Grade: ....

Learn 2 Think

200 Challenging Math Problems

every 6th grader should know

New edition 2012
Copyright Learn 2 Think Pte Ltd

email: contactus@learn-2-think.com

ISBN: 978-981-07-2767-3

Master Grade 6 Math Problems

Introduction:

Solving math problems is core to understanding math concepts. When Math problems are presented as real-life problems students get a chance to apply their Math knowledge and skills. Word problems progressively develop a student's ability to visualize and logically interpret Mathematical situations.

This book provides numerous opportunities to every student to practice their math skills and develop their confidence of being a lifelong problem solver. The multi-step problem solving exercises in the book involve several math concepts. Student will learn more from these problems solving exercises than doing ten worksheets on the same math concepts.

The book is divided into 9 chapters. Within each chapter questions move from simple to advance word problems pertaining to the topic. The last chapter of the book explains step wise solutions to all the problems to reinforce learning and understanding.

How to use the book:

Here is a suggested plan that will help you to crack every problem in this book and outside.

Follow these 4 steps and all the Math problems will be a NO PROBLEM!

Read the problem carefully:

- What do I need to find out?
- What math operation is needed to solve the problem? For example addition, subtraction, multiplication, division etc.
- What clues and information do I have?
- What are the key words like sum, difference, product, perimeter, area, etc.?
- Which is the non-essential information?

Decide a plan

- Develop a plan based on the information that you have to solve the problem. Consider various strategies of problem solving:
- Drawing a model or picture
- Making a list
- Looking for pattern
- Working backwards
- Guessing and checking
- Using logical reasoning

Solve the problem:

Carry out the plan using the Math operation or formula you choose to find the answer.

Check your answer

- Check if the answer looks reasonable
- Work the problem again with the answer
- Remember the units of measure with the answer such as feet, inches, meter etc.

Master Grade 6 Math Problems

Note to the Teachers and Parents:

✎ Help students become great problem solvers by modelling a systematic approach to solve problems. Display the 'Four step plan of problem solving' for students to refer to while working independently or in groups.

✎ Emphasise on some key points in the problem.

✎ Enable students to enjoy the process of problem solving rather than being too focused on finding the answers.

✎ Provide opportunities to the students to think; explain and interpret the problem.

✎ Lead the student or the group to come up with the right strategy to solve the problem.

✎ Discuss the importance of showing steps of their work and checking their answers.

✎ Explore more than one possible solution to the problems.

✎ Give a chance to the students to present their work.

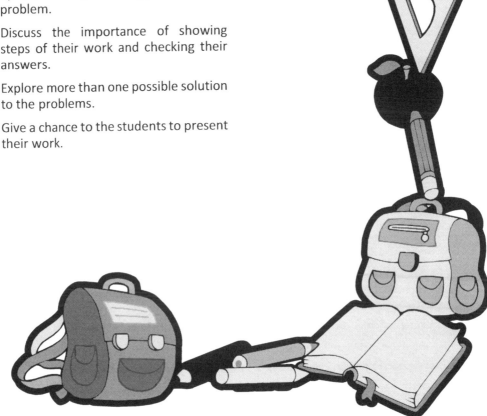

Contents

Topics   **Page number**

NUMBERS

Guaranteed to improve children's math and success at school.

Study the pattern in the table below.

1st Term	2nd Term	3rd Term	4th Term	5th Term
4	9	16	25	36

a) How can we relate the term to the
number below it?

b) What would be the 11th term?

Answer: ....

A book had pages numbered from 1 to 120. What is the sum of all the numbers written on every page?

Answer: ....

PROBLEM 3

Evaluate $12 + 20 \div 4 - 16 \times 2 + 32 \div 2$

Answer: ....

PROBLEM 4

What is the sum of all integers from 1 to 100?

Answer: ....

The sum of 5 consecutive whole numbers is 5920. What is the sum of all the digits of these 5 numbers?

Answer: ....

25 consecutive integers add up to 700.
What is the largest of these numbers?

Test 3
begins

Answer: ....

PROBLEM 7

The sum of the squares of the first 20 positive integers is 2870. What is the sum of the squares of the first 15 positive integers?

Answer: ....

PROBLEM 8

A number is ten times the second number. The sum of the two numbers is 5016. What is the difference between the two numbers?

Test !

Deepin

Answer: ………………………….

The product of two numbers is 48. Given that one number is 8 more than the other, find the two numbers.

Answer: ....

PROBLEM 10

Pete found the difference between two numbers and the result was 12. Michelle multiplied the same two numbers and the product was 540. What is the sum of the two numbers?

Answer: ....

PROBLEM 11

I have 2 numbers whose product is 78. The sum of the two numbers is a prime number. Both of my numbers are greater than the lowest even number. What numbers do I have?

Answer: ....

A palindrome is any work or number which reads the same forwards or backwards. For example, the number "12321" and the word "level" are both palindromes. How many whole numbers between 100 and 1000 are palindromes?

Answer: ....

PROBLEM 13

A teacher arranges 4 students A, B, C and D in a row? How many different ways of arranging them are possible?

Answer: ....

There are 3 clocks in Jesper's house. One chimes after every 4 minutes. The second clock chimes every 6 minutes. He also noticed that the third clock chimed at 10:40 A.M. and at 10:48 A.M. If all 3 clocks chimed at 2:00 P.M., when is the next time they will chime again?

Answer: ....

PROBLEM 15

Peter had 2 hourglasses. The sand in hourglass 1 runs out in 7 minutes and the sand in hourglass 2 runs out in 9 minutes. If Peter reset the hourglasses together at 2:57 P.M. when is the next time he should reset both the hourglasses at the same time?

Answer: ....

PROBLEM 16

The mean of the heights of five basketball players was 1.88 meters. When two new basketball players with the heights of 1.81 m and 1.93 meters were added, what was the new mean height of the basket ball players?

Answer: ....

PROBLEM 17

4 students could sit on a table, with one on each side. However if two tables were joined together only 6 students could sit at the table. How many tables would it take to seat 200 students, if they were all joined together?

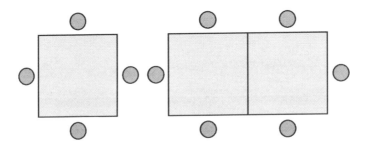

Answer: ....

PROBLEM 18

There were 4 marbles A, B, C and D. Marble C weighed 30 grams. Marble A and B weighed 120 grams combined together. If the total weight of four marbles was 170 grams; how much did Marble D weigh?

Answer: ....

PROBLEM 19

Carl was at a horizontal distance of 300 meters away from the Eiffel tower. If the distance from the top of the tower to Carl's head was 500 meters, what is the height of the Eiffel tower? (You may negate Carl's height)

Answer: ....

PROBLEM **20**

The heights of 6 basketball players are listed below.

1.72 m, 1.86 m, 1.84 m, 1.92 m, 1.78 m, 1.80 m

a) What was the average height of the players?

b) 4 years later, 3 of the players grew by 6 cm . What is the new average?

Answer: ....

The number of toffees in a container is between 50 and 100. If they are put into packets of 3, there will be 1 sweet left. If they are put into packets of 5, there will also be one sweet left. If they are put into packets of 7, there will be no sweets left. How many sweets are there in the container?

Answer: ....

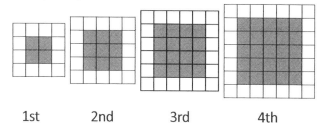

Study the pattern below.

1st   2nd   3rd   4th

How many white and colored squares are there in the 5th figure?

a) Create a formula to calculate the number of uncolored squares and the number of colored squares.

b) Calculate the number of colored and uncolored squares in the 11th figure.

Answer: ....

PROBLEM 23

The average weight of boxes A, B, C and D is 140 kilograms. This is 20 kilograms heavier than the average weight of boxes D, E and F. Box A is 20 kilograms heavier than box E. The average weight of boxes B, C, D is 160 kilograms and the weight of Box F is 120 kilograms.

a) What is the weight of box D?

b) What is the average weight of all these boxes?

Answer: ....

PROBLEM 24

Jim the squirrel feels hungry and needs to eat a nut every 6 minutes. His brother Joe feels hungry every 7.5 minutes and he also then goes and picks up a nut to eat. Their sister Julie feels hungry every 12 minutes and that's when she goes and eats a nut. If they ate a nut together at 2:30 P.M., at what time will they eat together again?

Answer: ....

A row of 9 juice cans are lined up in a row and 8 are stacked on top, then 7, then 6 and so forth until there is only 1 can stacked on the very top. How many juice cans were used to form this arrangement?

Answer: ....

PROBLEM 26

Two athletes A and B start running along a circular track. Athlete A runs 6 full circles in 120 minutes. Athlete B makes 5 full circles in 60 minutes. If they start running now from the same point, when will they be at the same starting point again?

Answer: ....

PROBLEM 27

There are 9 people at a party. They shake hands with each other at the beginning of the party. How many handshakes took place?

Answer: ....

Kathleen had an average score of 56 marks in the last three tests. How many marks must she score in the fourth test so that she can get an average of 65 marks?

Answer: ....

a) Simplify the following expression:

$$\frac{x - xy}{(1 - y)(1 + y)}$$

b) Use your answer in a) to find the solution to

$$\frac{8 - 48}{(-5) \times (7)}$$

Answer: ………………………

a) Find the HCF of 72 and 126.

b) A competition was organized. There were 72 girls and 126 boys. The instructors wanted teams to be formed whereby there were as few teams as possible. Each team had only 1 gender and there were an equal number of students in each team, How many teams were there?

Answer: ....

PROBLEM 31

The average of six numbers in a list is 64. The average of the first two numbers is 26. What is the average of the last four numbers?

Test 6 Deepening

Answer: ....

PROBLEM 32

A postman delivered 140 letters during the first hour on a certain day, 120 letters during the second hour and 170 letters during the third hour. How many letters must he deliver during the fourth hour in order to average 150 letters per hour for the four-hour period?

46 Test 5 keeping

Answer: ....

In a class there are 29 children. 12 children have a sister and 18 children have a brother. Tim, Ben and Lucy have no brother and no sister. How many children in that class have both a brother and a sister?

Answer: ....

ALGEBRA

6

Guaranteed to improve children's math and success at school.

PROBLEM 34

There are some Chinese, English and Spanish books on a shelf. $\frac{1}{4}$ of the books are Chinese books. $\frac{1}{5}$ of the remaining books are English. There are 63 more Spanish books than English books. How many books of each type are there?

Answer: ....

PROBLEM 35

Jane and John had some marbles. After Jane gave John some marbles, he had 3 times as many marbles as he had in the beginning. Jane then had half the number of marbles John had. If there were a total of 180 marbles, how many marbles were there in the beginning?

Algebra Word Problem 96

Answer: ....

Grace answered 15 questions in a test and received 29 points. If she got 3 points awarded for all her correct answers and 1 point deducted for each wrong answers, how many questions did she get wrong?

Answer: ....

Mrs Jones saved 10% of her monthly salary in January. In February, she saved 50% more than what she saved in January. She had $800 more after the second month than after the first month. What is the monthly salary of Mrs Jones?

Test 2
Deepin

Answer: ....

PROBLEM 38

$\frac{2}{3}$ of the class passed their Science examination. 80% more boys than girls passed their science examination. 4 more girls than boys did not pass the examination. If there were 42 students in the class, how many boys and girls were there?

Answer: ....

$\frac{1}{3}$ of the people at a party are men. When half the number of men leave the party, there were 192 more women then men. If 16 women leave the party, find the ratio of the number of women to the number of men left at the party.

Answer: ....

PROBLEM 40

Freddy and Justin were given a sum of money by their parents. If Freddy and Justin spend $8 and $16 daily, Justin would still have $70. If Freddy and Justin spend $16 and $8 daily, Justin would have $430 left. How much did their parents give each of them at first?

Answer: ....

PROBLEM 41

Three Identical tubs, A, B and C were filled with water. Tub A had a mass of 3.1 kilograms when it was half full. Tub B had a mass of 2.2 kilograms when it was one – fifth full. How full must Tub C be to have a mass of 2.8 kilograms?

Answer: ....

PROBLEM 42

2 apples and 1 orange costs $1.20. 2 oranges and 1 apple costs $1.80. Find the cost of an apple and an orange.

Answer: ....

PROBLEM 43

Adrian and Judy had some toy cars. They were then given an equal number of toy cars and Judy now had twice the number toy cars of what she had initially. $\frac{1}{3}$ of Judy's cars were now $\frac{1}{2}$ of Adrian's cars. If Judy had 36 more cars than Adrian,

a) How many toy cars were given to each of them?

b) How many toy cars did Adrian have at first?

Answer: ....

PROBLEM 44

In a class 20% are boys. 20% of the boys and 10% of the girls keep pets. If 144 girls do not keep pets, how many students are there in the class?

Answer: ....

PROBLEM 45

A group of not more than 40 pupils went for a trip. They paid a total of $364 for this trip. Each girl paid $10 and each boy paid a dollar less. If there were 6 more girls than boys, how many boys were there in the group?

Answer: ....

PROBLEM 46

A pet shop sells canaries, puppies and kittens. There are four times as many canaries as kittens in the shop. When the shop assistant makes a count of the animals, she find that there are 30 heads and 88 legs.

a) How many canaries are there in the shop?

b) How many puppies are there in the shop?

Answer: ....

PROBLEM 47

Mrs Underwood sold a total of 120 butter and chocolate chip cookies at a charity fair. Each butter cookie cost $2 and a chocolate chip cookie costs $4. She earned a total of $310. How many cookies of each type did she sell individually?

Answer: ....

PROBLEM 48

A farmer had twice as many chicken as roosters. After he sold 244 chickens, he had half as many chickens as roosters. How many roosters did he have?

Answer: ....

PROBLEM 49

A grocer sold 20% more apples than pears and 10% more pears than oranges. He sold 198 apples. If apples were sold in packets of 6 for $2, pears were sold in packets of 11 for $10, and oranges were sold in packets of 10 for $3, how much did the grocer earn?

Answer: ....

PROBLEM 50

Mark had 4100 beads more than his sister. After Mark gave 900 beads to his sister, he had thrice as many beads as his sister. How many beads did Mark have at first?

Answer: ....

Colin gave 20% of his salary to his wife and spent 30% of the remainder. Then he gave the rest of the money to sons Jason and James in the ratio 5:3. How much percentage of his money did Jason receive from Colin?

Answer: ....

PROBLEM 52

In a book store, there were 1647 more English books than Chinese books. When 2400 English books were sold out, there were 4 times as many Chinese books than English books left in the book store. Find the total number of books that were in the book store at first.

Answer: ....

PROBLEM 53

Janice has some savings in the bank. She decided to spend 25% of her savings on clothes and the remaining amount on a television set and a Wi-Fi set in the ratio 2:3. She found that she had spent $280 more on the hi-fi set than on clothes. How much was her original savings?

Answer: ....

PROBLEM 54

Marcus, Fred and Janice shared the cost of a dinner. Marcus paid thrice as much as Fred. Fred paid twice as much as Janice. If Marcus paid $25 more than Janice, what was the cost of the dinner?

Answer: ....

PROBLEM 55

Terry and Sarah had $745 altogether. After their mother gave Sarah another $15, Terry had thrice as much money as Sarah. How much more money than Sarah did Terry have at first?

Answer: ....

PROBLEM 56

Sam's savings are 10% more than Thomas. If Sam transfers $240 from his savings account to Thomas's account, they will have the same amount of savings. How much savings does Sam have?

Answer: ....

PROBLEM 57

A minute ago, 80% of the pupils in a hall were girls. 140 more pupils entered the hall. As a result, the number of girls increased by 25% and the number of boys increased by 75%. How many pupils are there in the hall now?

Answer: ....

PROBLEM 58

Andy has 160 kilograms more rice than flour. After selling 400 kilograms of rice and 30 kilograms of flour, he has thrice as much flour than rice. What was the total mass of rice and flour he had at first?

Answer: ....

PROBLEM 59

Gavin took a test with 30 questions. For every question he got right, he earned 10 points, and for every question he got wrong, he lost 2 points. He answered every question, and got a score of 120. How many questions did he get right?

Answer: ....

PROBLEM 60

Daniel has $60 in his piggy bank. There were a mixture of 20 cent and 50 cent coins. There were 8 more 50 cent coins than 20 cent coins. What is the total number of coins Daniel has in his piggy bank?

Answer: ....

PROBLEM **61**

A pen costs $7 more than a wallet and a tie costs $5 less than the wallet. If the pen and the tie cost $21 altogether, find the cost of the wallet.

Answer: ....

PROBLEM 62

Alex has $1.50 more than Betty and three times as much money as Colin. The 3 of them have $11.80 altogether. How much money does Colin have?

Answer: ....

PROBLEM 63

The sum of the age of Allan and Bernard is 20. The sum of the age of Bernard and Cindy is 21. The sum of the age of Cindy and Allan is 25. How old is Allan?

Test 6 Requis

Answer: ....

PROBLEM 64

Eric had 46 ten-cent, twenty-cent and fifty-cent coins which add up to $16.10. There were 9 ten-cent coins and the rest were twenty-cent coins and fifty-cent coins. How many 20-cent and 50-cent coins were there?

Answer: ....

Joe, Lucy and Mark are members of the same family. Lucy is 10 years older than Joe. Mark is 7 years older than Lucy. The sum of their ages is 45 years. How old is each one them?

Answer: ....

PROBLEM 66

A school has 10 classes with the same number of students in each class. One day, many students were absent. 5 classes were half full, 3 classes were 3/4 full and 2 classes were 1/8 empty. A total of 50 students were absent. How many students are in this school when no students are absent?

Answer: ....

PROBLEM 67

Samantha went shopping. She bought the following fruits: apples, oranges and peaches. The apples and oranges weigh 15 kg together. The oranges and peaches weigh 18 kg together. The apples and peaches weigh 16 kg together. How many kg of fruits did Samantha buy altogether?

Answer: ....

PROBLEM 68

A piece of wire is 120 centimeters long.
It is cut into 3 parts. The longest piece is
twice the length of the second part and the
shortest part is 21 centimeters long. How
long is the longest piece of the wire?

Answer: ....

PROBLEM 69

Total number of students in school A is 30% of that of school B. The number of students in school B is 60% that of school C. If 1800 students were transferred from school C to school A, the number of students in school C will be 1/6 of the number of students in school B. Find the total number of students in the 3 schools.

Answer: ....

PROBLEM 70

There was 0.6 Litres of water in a bottle and 4.2 Litres of water in a jug. I poured an equal amount of water into each container. Now the amount of water in the jug is 5 times the amount of water in the bottle.

a) How much water did I pour into each container?

b) Find the total amount of water in the 2 containers now.

Answer: ....

PROBLEM 71

Emily had a ribbon with her. She gave one third of it to her sister and used 40% of the remaining on her dress. She cut the remaining ribbon into four equal parts and gave to her four cousins. If each cousin got 15 cms of the ribbon, what was the length of the ribbon used on the dress?

46 Test 5 keeping

Answer: ....

PROBLEM 72

A jar filled with 40 identical screws weighs
1.4 kilograms. The same jar when filled with
20 identical nails weighs 500 grams. The
mass of each screw is twice the mass of
each nail. What is the mass of the empty jar.

Answer: ....

PROBLEM 73

The only animals on Farmer Justin's farm are cows and chickens. If Justin counts a total of 45 heads and a total of 126 feet, find the number of cows on Justin's farm.

Answer: ....

PROBLEM 74

Richard took My Math Olympics test with 25 questions. For every question he got right, he earned 10 points, and for every question he got wrong, he lost 2.5 points. He answered every question, and got a score of 150. How many questions did he get right?

Answer: ....

PROBLEM 75

40% of adults left a party before it ended. 10% of those adults are men. The ratio of men who left to those who did not is 2 : 5. If there were 105 men at the party in the beginning, how many women were there in the beginning?

Answer: ....

PROBLEM **76**

Sandra had some oranges and bananas. After selling 1/3 of the oranges and 7 bananas, the number of bananas left was a quarter of the number of oranges left. If the number of bananas sold was 1/7 of the original number of bananas, how many of each fruit were there at first?

Answer: ....

FRACTIONS, DECIMAL AND PERCENTAGES

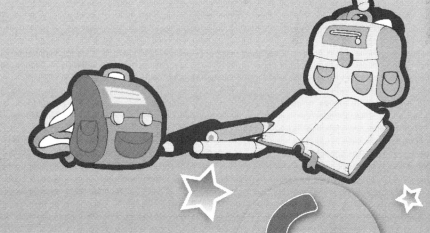

Guaranteed to improve children's Math and success at school.

PROBLEM 77

The science test Charles took this week had 45 questions. For each correct answer Charles scored 4 marks. He answered 85% of the questions correctly. Last week, Charles had taken the same test but he could answer only 65% of the questions correctly.

How many more questions did Charles answer correctly on this week's test than on last week's test? How much did he score in each test?

Answer: ....

PROBLEM 78

Tommy had some green and yellow beads. 40% of his beads were green. When he lost 50 yellow beads, the number of yellow beads fell to $\frac{2}{3}$ of the initial amount of yellow beads. How many beads did Tommy have in the end?

Done

Answer: ....

PROBLEM 79

Sam had a total of $7.20 of 5 cent and 20 cent coins. $\frac{2}{3}$ of the coins were 20 cent coins. How many coins of each type did he have?

Answer: ....

$\frac{2}{3}$ of a number is greater than $\frac{1}{2}$ of the same number by 12. What is the difference between the number and its 3rd multiple?

46 Tot 5 degeng

Answer: ....

There were a total of 2250 seats in a theatre. 10% of the seats were first class, 30% of the seats were second class and the rest of the seats were third class. 2 years later, 100 first class seats and 125 second class seats were added.

a) How many first class seats were there in the end?

b) How many second class seats were there in the end?

c) What percentage of seats were third class in the end?

Test Begins

Answer: ....

847 runners participated in a 5 kilometres running contest. 3/5 of the female participants were girls and 1/4 of the male participants were boys. There were thrice as many men as women. How many adults participated in the running contest?

Answer: …………………………

PROBLEM 83

Charles and Ryan had a total of 432 game cards. Charles gave $\frac{2}{5}$ of his game cards to Ryan. Then, Ryan gave $\frac{1}{4}$ of the total number of game cards he had to Charles. Both of them had the same number of game cards in the end.

How many game cards did Charles have at first?

Answer: ....

PROBLEM 84

In a concert $\frac{1}{6}$ of the audience were adults. 60% of the children were girls. There were 48 more girls than adults.

a) How many girls were at the concert?

b) Half way through the performance, some boys left the concert. As a result, only 25% of the remaining audience were boys. How many boys left half way through the concert?

Deepis
Test 3

Answer: ....

PROBLEM 85

Carmen bought some red and blue files at a bookshop. $\frac{2}{3}$ of the files she bought were red and the rest were blue. Later, she gave away $\frac{3}{4}$ of the red files and $\frac{1}{4}$ of the blue files and had 100 files left. How many files did she buy at first?

Done (45)

Answer: ....

PROBLEM 86

Harry has 60 postcards. He gave 0.25 of them to his niece and $\dfrac{3}{5}$ of the remainder to his cousin. How many post cards were left?

Answer: ....

PROBLEM 87

During a shopping spree, Rachel spent $\frac{1}{5}$ of her money on a skirt and $\frac{1}{3}$ of the remaining money on 4 blouses and a belt. She spent the rest of her money on a bottle of perfume. The belt cost $\frac{1}{4}$ as much as the bottle of perfume. If she spent $21.50 on each blouse how much did she pay for the bottle of perfume?

Answer: ....

PROBLEM 88

Jenny uses 25 liters of water and 15 liters of orange cordial to make orange punch.

a) What percentage of the orange punch is the orange cordial?

b) How many more liters of orange cordial must she use to get about 50% of orange cordial in the mixture?

Done

Answer: ....

PROBLEM 89

Hilda had a $3\frac{1}{5}$ meters long wire. She cut it into 8 equal pieces and used 5 pieces to make a lantern.

a) Find the length of each piece of wire.

b) Find the total length of the wire used to make the lantern. Express the answer in its simplest form.

Answer: ....

PROBLEM 90

Kathy's weight is $\frac{5}{7}$ of Harry's weight.
Tom's weight is $\frac{3}{5}$ of Kathy's weight. Harry
is 37.2 kilograms heavier than Tom.

Find Harry's weight.

Done

Answer: ....

PROBLEM 91

$\frac{2}{5}$ of the people who attended the Paris World Expo were Chinese. There were 32 more French than Chinese. The remaining 58 were Dutch. How many people were there at the Paris World Expo?

Done 91

Answer: ....

PROBLEM 92

$\frac{2}{3}$ of the mangoes in the basket were overripe and the rest were ripe. When 40 mangoes were taken out of the basket, $\frac{3}{4}$ of the remaining mangoes in the basket were overripe. If there were a total of 120 mangoes in the basket at first,

a) How many overripe mangoes were taken out of the basket?

b) How many ripe mangoes were taken out from the basket?

c) How many more overripe mangoes than ripe ones were left in the basket?

Answer: ....

PROBLEM 93

600 pupils were asked to choose their favorite destination. 26% of them chose Thailand, 20% chose London, 12% chose Italy and the rest chose Hong Kong. How many more pupils choose Hong Kong over London?

Answer: ....

PROBLEM 94

Kelvin had 1200 apples. 5% of them were rotten and thrown away. 40% of the remainder were packed into boxes of 12 apples each and the rest into large boxes of 36 apples each. If he sold the small boxes of apples at $4.30 per box and the large boxes at $13 per box, how much money did he collect from his total sales?

Answer: ....

PROBLEM 95

A squash club had a certain number of male and female members. When 480 females left, the membership was decreased to 80% of its original enrolment. How many members were there in the club initially?

Answer: ....

PROBLEM 96

A fruit seller had a certain number of apples. He sold $\frac{1}{5}$ of them to Lena and 35% of them to Caroline. He then found out that out of the remaining apples, 10 of them were rotten and he threw them away. The number of apples sold was 30 more than the number of apples left. How many apples did he have at first?

Answer: ....

PROBLEM 97

There were 200 pupils in a lecture theatre. 30% of them were girls. When some girls left the theatre, the percentage of girls in the theatre dropped to 20%. How many girls left the theatre?

Done

Answer: ....

PROBLEM 98

James sold 25% of his tarts on Monday and 80 fewer tarts on Tuesday than on Monday. On Wednesday, he sold 160 tarts and found that he had 30% of his tarts left. How many tarts did he sell in all?

Done

Answer: ....

PROBLEM 99

Rachael withdrew a sum of money from the bank. She spent half of the sum of money on a wardrobe, $360 on a dressing table and had $\frac{1}{8}$ of the sum of money left. How much did she spend on the wardrobe and the dressing table?

Answer: ....

PROBLEM 100

There are some marbles in a bag. Bob takes 1/5 of them for himself. John takes 3/4 of the remaining marbles. Then Tom takes the remaining four marbles. How many marbles were there in the beginning?

Answer: ....

PROBLEM 101

40% of a Karate class of 40 are boys. If some girls joined the class and the percentage of the girls increased to 80%, how many new girls joined the class?

Test 4
Deepti

Answer: ....

PROBLEM 102

A sum of money was shared among 3 children. James got 20% of the money and Alice got $2800. William got 60% of the money Alice got. Find the sum of money, that was shared among 3 children.

Answer: ....

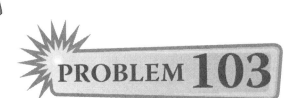

PROBLEM 103

Penny and Andy spent $1980 altogether.
Penny spent 20% less than Andy. How much
did Andy spend?

Test 6 papers

Answer: ....

Learn 2 Think

RATIO AND PROPORTION

6

Guaranteed to improve children's math and success at school.

PROBLEM 104

The model of a building is made to the scale 2:450.

If the actual building is 81 meters high, how tall will the model of the building be in centimeters?

Answer: ....

There are 32 green markers and 5 times as many red markers in the box. How many more green markers must be added in the box so that the ratio of the number of green markers to red markers is 3 : 4?

Ratio Problems
Yb
Adapted
↓
Y6 Test 6
accepted

Answer: ………………………

PROBLEM 106

A cheesecake costs $30 and a mango cake costs $36. Last month the total amount of money collected from the sale of the 2 types of cakes was $3060. For every 15 cakes ordered, 5 were cheesecakes and 10 were mango cakes.

a) What was the total number of cheesecakes sold?

b) How much was obtained from the sales of mango cakes?

RatioProblems 4/6

Answer: ....

PROBLEM 107

Mark thinks of two numbers. The first number is 3/8 of the second number. If the difference between the two numbers is 45, find the value of the second number.

*More ratio
problems 76*

Answer: ....

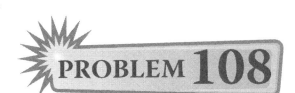

PROBLEM 108

The ratio of the number of men to the number of women in a party is 3:7. There are 36 more women than men. How many people are there at the party?

*Mixe Ratio
Problems 46*

Answer: ....

PROBLEM 109

18 years ago, the ratio of Carl's age to David's age was 7:2. Now the ratio of Carl's age to David's age is 2:1.

a) What is Carl's age now?

b) Find David's age 18 Years from now.

Answer: ....

PROBLEM 110

Nancy collected $\frac{1}{4}$ as many shells as William. William collected 40 shells more than Richard. Altogether, William and Nancy collected 70 more shells than Richard. How many shells did William collect?

Answer: ....

At a party, the number of Indians present was 25% of the number of Chinese. After 24 Chinese and 13 Indians joined the party, the ratio of the number of Chinese to the number of Indians was 3:1.

How many Chinese were at the party at first?

Answer: ....

PROBLEM 112

The ratio of James's savings to Margaret's savings was 9:2. After James spent $80 and Margaret spent $30, the ratio became 10:1.

a) How much money did James have at first?

b) How much money did Margaret have after spending $30?

Answer: ....

PROBLEM 113

The ratio of adults to children at a gathering was 4:5. When 6 children left, the ratio of adults to children became 8:7. How many children were there at first?

Answer: ....

PROBLEM 114

Initially, there were a total of 180 beads. The ratio of the number of blue beads to the number of red beads in a container was 5:4. When some blue beads and red beads were added to the container, the ratio became 3:2. The total number of beads became 300.

a) How many blue beads were added?

b) How many more blue beads than red bedas were added?

Answer: ....

PROBLEM 115

Lisa made Kiwi tarts, apple tarts and peach tarts in the ratio 2:3:5. After she made 12 new tarts of each kind, the ratio became 7:9:13. Find the number of apple tarts she has now.

Answer: ....

PROBLEM 116

There are some apples in 3 baskets, A, B and C. The ratio of the number of apples in A to B is 3:4 .There are 98 apples in C. If the ratio of the total number of apples in A and B to the number of apples in C is 4:2, how many apples are there in box basket A?

WS Ratio beguin 96

Answer: ....

PROBLEM 117

At the start of the week a bookshop had story books and art books in the ratio 6 : 2. By the end of the week, 25% of each type of books were sold and 1620 books of both types were unsold. How many books of each type were there at the start of the week?

W16 Portion Requij 76

Answer: ....

PROBLEM 118

Rachel and John had a total of 80 marbles in the ratio 2 : 3. When John lost some marbles, the marbles that he had remaining made up $\frac{1}{3}$ of the total number of marbles they both had left. How many marbles did John have in the end?

Answer: ....

PROBLEM 119

The ratio of cars to vans in a car lot was 1 : 2. When 68 cars entered the carpark, and 18 vans left the carpark, the ratio of cars to vans became 7 : 3. How many cars and vans were there in the carpark at first?

Answer: ....

AREA, PERIMETER AND VOLUME

6

Guaranteed to improve children's math and success at school.

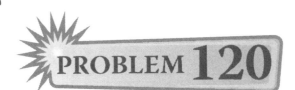

Two circle shared the same centre. The first circle had a radius of 7 meters and the second had a radius of 14 meters. What was the ratio of the areas of the smaller circle to the bigger circle?

Answer: ....

132

PROBLEM 121

A rectangular tank 65 cm long, 40 cm wide and 35 cm high was 1/2 filled with water. The water was then used to completely fill up some bottles. The capacity of each bottle was 1.5 liters. How many bottles were filled completely? What was the amount of water left in the tank when all the bottles were filled completely?

Answer: ....

PROBLEM 122

Florence has rectangular blocks that are 5 cm by 4 cm by 2 cm. What is the greatest number of these blocks that can fit in a cubic box whose inner dimensions are 10 cm by 12 cm by 10 cm?

Answer: ....

PROBLEM 123

The height of a rectangular tank is 10 m.
This tank contains 720 m^3 of water when
filled to the 4 m mark. What is the capacity
of the tank?

Test 2
Required

Answer: ....

PROBLEM 124

A water tank, 10 m high with a square base of 4 m, is full of water. The water is poured into another tank 8 m long and 4 m wide. What is the height of the water in the second tank?

Answer: ....

PROBLEM 125

Each of the curved lines form a quater of a circle. Find the perimeter of the following figure. Take π as $\frac{22}{7}$

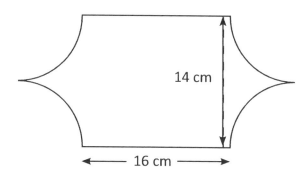

14 cm

16 cm

Answer: ....

PROBLEM 126

What is the sum of all the internal angles in the shape below?

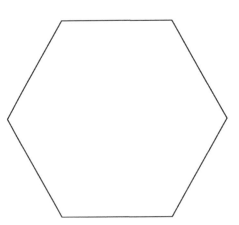

Answer: ....

PROBLEM 127

Find the area of the shaded part, given that the radius is 28 cm. Take $\pi = \dfrac{22}{7}$

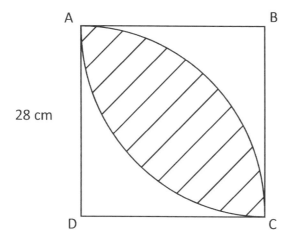

28 cm

Answer: ....

PROBLEM 128

The perimeter of a rectangle is twice that of a triangle below. How many possible values of its length are there if each length is a whole number?

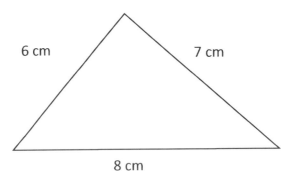

6 cm

7 cm

8 cm

Answer: ....

PROBLEM 129

A rectangular tank measuring 14 cm by 20 cm by 42 cm was 3/4 filled with water. The water was then released through a hole at a rate of 1 liter per minute. After 5 minutes, what fraction of the tank was not filled with water?

Answer: ....

PROBLEM 130

Four equal rectangular wooden planks are joined together to make a square photo frame. Each rectangle has a perimeter of 40 cm. What is the area of the square photo frame ABCD?

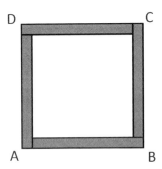

Answer: ....

142

A rectangular prism is made up of cubes of sides 1 cm each as shown below.

What is the volume of this rectangular prism?

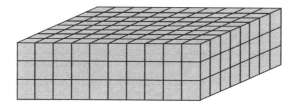

Answer: ....

PROBLEM 132

There is a rectangular box measuring
10 centimeters by 6 centimeters by 4
centimeters. How many cubes of side 2
centimeters each can you put into this box?

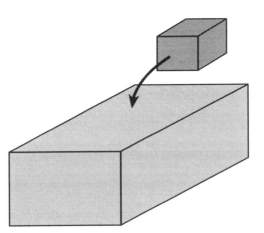

Answer: ....

144

PROBLEM **133**

¾ of a rectangular glass tank which is 60 cm long, 40 cm wide and 25 cm high is filled with water. Ben transfers some of the water from the glass tank to a steel tank until it is filled to its brim. 15 liters of water is left in the glass tank. The steel tank is 25 cm long and 20 cm wide. Find the height of the steel tank.

Answer: ....

PROBLEM 134

4 quadrants of equal radius were cut from a square cardboard of sides 28 cm.

Find:

a) The perimeter of the cut cardboard

b) The area of the shaded region.

   (take $\pi = \dfrac{22}{7}$)

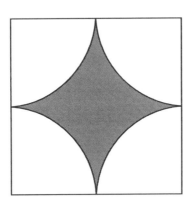

Answer: ....

PROBLEM 135

The figure shows 4 semi circles and a rectangle. The length of rectangle is twice its width. If the length of the rectangle is 14 cm, Find the perimeter and area of the figure . (take $\pi = \frac{22}{7}$)

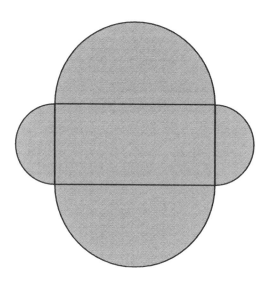

Answer: ....

PROBLEM 136

The length of a rectangle is 140% of its width. Find the length and width if the perimeter of the rectangle is 144 cm.

46 Test 5 eequis

Answer: ....

Given that ABCD is a square, find the area of the 4 "leaves".

(take $\pi = \dfrac{22}{7}$)

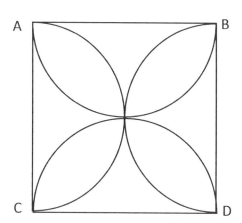

Answer: ....

PROBLEM 138

A rectangular tank, 90 cm by 50 cm by 30 cm, was at first 30% filled with oil.

A tap flowing at a rate of 900 cm^3 per minute was turned on. How long did it take the tap to fill the tank to its brim?

Answer: ....

PROBLEM 139

A tank was being filled with water from two taps simultaneously at a rate of 6 liters a minute from each tap. At the same time the water was being drained out of a hole at the bottom of the tank at a rate of 8 liters every 4 minutes. If the tank was only 40% full after 10 minutes,

a) What was the capacity of the tank ?

b) How much longer will it take to fill the tank?

Answer: ....

PROBLEM 140

60% of a piece of land was used to construct a school. 1/4 of the remaining land was used to construct a garden. The perimeter of the garden was 78 meters. If its length was 24 meters;

Find the area of the original piece of land.

How much bigger was the area used to build the school than the garden?

Answer: ....

PROBLEM 141

A rectangular tank of base area 360 cm^2 was filled with water. When 1200 cm^3 of water is poured away, the remaining water is poured into 2 cm cubes. How many cubes are needed to pour the remaining water?

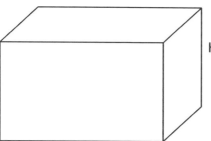

H = 15 cm

Answer: ....

PROBLEM 142

When Grace poured out 1.62 liters of water from a rectangular can, the volume of water was reduced by $\frac{1}{5}$. If the can had a base perimeter of 0.6 meters and its length exceeded its width by 6 cm, find the depth of water in the can if it was filled to its capacity originally.

Answer: ....

154

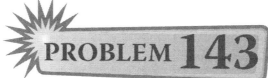

In the figure below, the diameter of each of the smaller circles is the radius of the larger circle. Find the shaded area of the figure. (take $\pi = \dfrac{22}{7}$)

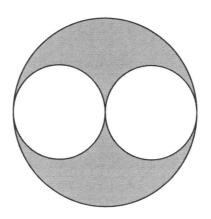

Answer: ....

A rectangular container measures 24 cm by 22 cm by 19 cm.

a) How many small cuboids of sides 4 cm by 3 cm by 2 cm can be neatly arranged in the container?

b) Will the cubes fill the whole container exactly? If not, what volume of the container will remain empty?

Answer: ....

If the ratio of the volume of cube A to cube B is 2:1, find the length of Cube B, when the volume of cube A is 512 cm^3.

432

Answer: ....

PROBLEM 146

Apple juice was allowed to flow out steadily from a tank full of apple juice. At the end of 6 minutes, the tank was 2/3 full. After further 8 minutes, the volume of apple juice left was 2.4 L.

a) Find the volume of the tank.

b) Find the volume of apple juice that flowed from the tank in 1 minute.

Answer: ....

Thomas had a pin, a string and a pencil. If the length of the string is 70 mm long, what is the area of the largest shape he can draw using the 3 objects? Leave your answer in cm^2.

Answer: ....

Below are the names of shapes and the sum of their interior angles.

Shape	Triangle	Rectangle	Pentagon	Hexagon
Sides	3	4	5	6
Sum of Interior Angles (degrees)	180	360	540	720

a) Form an equation relating the number of sides of the shape with the sum of it's interior angles.

b) Find the sum of interior angles of a decagon (10 sides).

Answer: ....

PROBLEM 149

A tank measures 15 cm by 20 cm by 30 cm. It was already $\frac{1}{5}$ filled with water. It was then further filled with water from 2 taps, simultaneously. Water flowed from the first tap at a rate of 60 ml/min and the second tap at a rate of 90 ml/min. How much faster will it take to fill the tank if both taps fill the tank at a rate of 100 ml/min?

Answer: ....

Learn 2 Think

SPEED, DISTANCE AND TIME

6

Guaranteed to improve children's math and success at school.

PROBLEM 150

A car took 5 hours to travel from Singapore to Kuala Lumpur. A bus took 9 hours to travel the same journey. If the car traveled at 90 km/hour,

a) What is the distance from Singapore to Kuala Lumpur?

b) What was the speed of the bus?

c) If the bus wanted to reach Kuala Lumpur 1 hour and 36 minutes earlier, and can only adjust its speed after $\frac{3}{5}$ of the journey how much faster should it travel?

Answer: ....

PROBLEM 151

Francis starts jogging at 2:15 P.M. along a path at 8 km/h. Christopher starts jogging at 2:30 P.M. along the same path. At 3:15 P.M. they both meet.

a) What is the distance that both of them jogged?

b) What is Christopher's jogging speed?

c) They both continue jogging further from 3:15 P.M. onwards. How long will it take for Christopher to be 6 kilometers ahead of Francis?

Answer: ....

PROBLEM 152

Shawn drove at a uniform speed of 90km/h. He started driving at 5:30 A.M. and reached his destination at 1:30 P.M. Mary started driving at the same time and reached her destination 6 hours later. What is Mary's speed?

Answer: ....

In one day, Johnson can make 360 kites and Debby can $\frac{2}{3}$ make as many kites as Johnson. How many days will both of them take to make 3600 kites.

Answer: ....

A carpenter took 6 minutes to cut a piece of wood into 2 pieces.

a) How long did he take to cut a piece of wood into 5 pieces?

b) How long did he take to cut a piece of wood into 30 pieces?

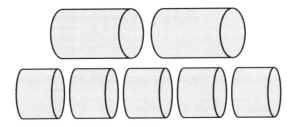

Answer: ....

PROBLEM **155**

Motorist A left Town X for Y at 9:30 A.M., traveling at a speed of 80 km/h. He arrived at town Y at 3:30 P.M. Motorist B left Town X 2 hours after Motorist A and he traveled at 100 km/h.

a) What time did motorist B reach Town Y?

b) How fast must motorist B travel if he were to reach Town Y at the same time as motorist A?

Answer: ....

Jason drove at 80 km/h for the first $\frac{5}{9}$ of his journey. He then completed the remaining 24 km of his journey in 12 minutes.

a) How long did he take to complete the whole journey?

b) What was the speed for the second part of his journey?

Answer: ....

Sam drove at an average speed of 84 km/h for 45 min. He then reduced his average speed by 6 km/h and drove further for 30 min before he reached his destination. Find the total distance covered by Sam.

Answer: ....

PROBLEM 158

At what time will car A overtake car B if car B left a town at 8.00 A.M. traveling at 60 kilometers per hour and car A left the same town at 11 A.M. traveling at 80 kilometers per hour and on the same path as car B?

Answer: ....

PROBLEM 159

At 10:30 A.M., a cyclist started traveling on a road at an average speed of 60 km/h. At 2:30 P.M., a motorist started from the same place, traveling on the same road. If the motorist took 4 hours to catch up with the cyclist, find his average speed.

Answer: ....

PROBLEM 160

The distance between Bangkok and Kuala Lumpur was 720 km. Andy took 6 hours to travel from Bangkok to Kuala Lumpur. Gavin left Bangkok 30 min earlier than Andy and traveled at 20 km/hour slower than Andy. How far away from Kuala Lumpur was Gavin when Andy reached Kuala Lumpur?

Answer: ....

PROBLEM 161

A motorist set off at 9:15 A.M. from Marina Bay to Anderson Point at an average speed of 80 km/h and arrived at 1:45 P.M. A lorry driver left Marina Bay 30 minutes earlier but reached Anderson Point 1 hour later than the motorist.

a) What was the average speed of the lorry driver?

b) How much speed should the lorry driver increase if he wanted to reach Anderson Point at the same time as the motorist?

Answer: ....

PROBLEM 162

The distance between Bay Front and New Town is 468 km. Kate left Bay Front at an average speed of 72 km/h. She maintained at this speed for 2 hours before reducing it by 18 km/h for the rest of the journey to New Town. What was her average speed from Bay Front to New Town?

Answer: ....

PROBLEM 163

Every Sunday, William goes to school which is 36 km away from his home. He always drives at an average speed of 48 km/h on his motorcycle. Last Sunday, he left his home 9 minutes earlier and still arrived at the school at the usual time. By how much did he decrease the speed of his motorcycle?

Answer: ....

PROBLEM 164

Victoria Central and Pebble Bay are 900 km apart. Margaret traveled at an average speed of 80 km/h from Victoria Central to Pebble Bay. At the same time, Alicia traveled at 70 km/h from Pebble Bay to Victoria Central. What is the distance covered by Margaret when they meet?

Victoria Central	Pebble Bay
X	900 – X

Answer: ....

It takes 40 workers 8 days to paint a building. How many workers are required to paint a building in 20 days?

Answer: ....

PROBLEM 166

A train traveled 900 km from Port of Eden to Port Stephens at an average speed of 150 km/h. It then moved on to Newcastle Port which is 600 km away at an average speed of 120 km/h. Find the time taken for the whole journey.

Answer: ....

PROBLEM 167

A Tiger Airways plane left Tokyo for the US and flew north at an average speed of 324 km/h. A Singapore Airlines plane left Tokyo an hour later but traveled at 360 km/h. How long did Tiger Airways travel before Singapore airlines caught up?

Answer: ....

PROBLEM 168

Jason and Gavin started driving from the same place but in opposite directions along a straight road. After driving for 2 hrs, they were 260 km apart. Jason's average driving speed was 60km/h. What was Gavin's average driving speed?

Answer: ....

PROBLEM 169

A round trip to Beijing is twice as long as a round trip to Sydney but half as long as a round trip to London. If the three round trips take 8 weeks to complete, how long is a one-way trip to Beijing?

46 Test s Beijing

Answer: ....

PROBLEM 170

At 10:30 A.M., a cyclist started traveling on a road at an average speed of 60 km/h. At 2:30 P.M., a motorist started from the same place, traveling on the same road. If the motorist took 4 hours to catch up with the cyclist, find his average speed.

Answer: ....

PROBLEM 171

Josh left Town A at 6:45 A.M. and drove at a uniform speed of 60 km/h to Town B. Caroline left Town A at 7:15 A.M. and drove towards Town B along the same expressway as Josh. She passed him at 10:15 A.M. and continued driving at the same speed until she reached Town B. If the distance between Town A and Town B is 560 kms, how far was Josh from Town B when Caroline reached Town B?

Answer: ....

PROBLEM 172

It takes 2000 bees one year to make 7 jars of honey. How many years will it take 5000 bees to make 70 jars of honey?

Answer: ....

An earthworm is at the bottom of a 6 meter well. Each day the earthworm walks up 3.5 meters but at night it slips down 2 meters. On which day will the earthworm get out of the well?

Answer: ....

PROBLEM 174

An aero plane flew from London to Amsterdam at 300 km/hr. It flew back from Amsterdam to London at 600 km/hr traveling the same distance each way. What was its average speed for the round trip?

Answer: ....

PROBLEM 175

Katie and Jim started walking to school at 8:00 A.M. at 4 kilometers per hour. 10 minutes later, Katie realized that she left her homework notebook at home. She told her brother Jim to run and bring it to her. If Jim ran at a speed of 8 kilometers per hour to get the notebook, and Katie kept walking towards the school at the same speed, at what time will Jim catch up with his sister?

Answer: ....

PROBLEM 176

My dog was 1000 meters away from home, and my cat was 800 meters away from home. When I called they both ran towards home. If my dog ran twice as fast as my cat, how far from the home was my cat when my dog reached home?

Answer: ....

PROBLEM 177

A bus and a car leave the same place and traveled in opposite directions. If the bus is traveling at 50 kilometers per hour and the car is traveling at 55 kilometers per hour. In how many hours will they be 200 kilometers apart?

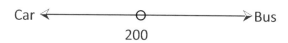

Car ⟵————————O————————⟶ Bus
200

Answer: ....

TIME, AGE AND MONEY

6

Guaranteed to improve children's math and success at school.

PROBLEM 178

Ms. Rachel is five times as old as her son Ben now. In 8 years time, their total ages will be 58 years. Kevin is twice as old as Ben now.

a) What is Ben's present age?

b) What is Ms. Rachel and Kevin's present age?

Answer: …………………………

Nicolas is 4 years old and Susan is 24 years old. When will Susan be thrice as old as Nicolas?

Tert 2
Deepens

Answer: ....

PROBLEM 180

When Gavin was 15 years old, his sister was 8 years old and his father was 45 years old. How old will Gavin's sister be when Gavin is half his father's age?

Test & sleeping

Answer: ....

PROBLEM 181

The sum of the ages of 5 siblings born at an interval of 3 years each, is 70.

What is the age of the youngest sibling?

Answer: ....

PROBLEM 182

Jim turned 35 years old this year. Three years ago, Jim was four times as old as his daughter Claire was then. How old is Claire this year?

Answer: ....

PROBLEM 183

Mr. Smith is 8 times the age of his grandson James. When James was born, his father was 3 times as old as James is today. If Mr. Smith was 32 years old when his son was born, how old is James today?

Answer: …………………………

PROBLEM 184

Fenny is $\frac{2}{7}$ of her fathers age in 2004. If her father was 33 years old in 1995, in which year will Fenny's age be $\frac{5}{11}$ of her fathers age?

Answer: ....

PROBLEM 185

In 1998 Mary's father was 4 times as old as her. Her brother was twice her age.

If Mary's Father was 36 years old?

a) How old will Mary's brother in the year 2000?

b) What was the total age of the family in 2003?

Answer: ....

PROBLEM 186

A brand new Sony television cost $1500. Carl paid 20% of the cost of the television as down payment and the rest of the cost in 12 monthly installments.

How much did Carl pay as down payment and how much as his monthly payment?

Answer: ....

PROBLEM 187

At an IT fair, a computer was sold at $1280 after a 20% discount. When Susan bought the computer, she was given a further 15% discount on the original selling price. How much more would she need to pay if she had bought the computer without discounts?

Answer: ....

PROBLEM 188

The price of a furniture in January was $1500. The price was increased by 12% in February. In March, it was decreased by 30%.

What was the difference in the price of furniture in January and in March?

Test 3 keepis

Answer: ....

PROBLEM 189

At a sale, Jacob reduced the price of all his products by 20%. Laura bought a camera and a laptop for $3520. If the original price of the camera was $900, how much did she pay for the laptop at the sale?

Word problems Multiply %

Answer: ....

PROBLEM 190

Michael bought a total of 60 kittens and turtles. Each kitten cost $5. Each turtle cost $3. If the total cost of the kittens was $100 more than the cost of turtles;

a) How many turtles did Michael buy?

b) Find the total cost of the kittens he bought?

Answer: ....

PROBLEM 191

A clock that uniformly loses 8 minutes every 24 hours was correctly set at 6 A.M. on January 1. What was the time indicated by this clock when the correct time was 12 o'clock noon on January of 5th of the same year?

Answer: ....

PROBLEM 192

Molly wanted to buy a new hand bag. She had only 1/2 of the cost of the hand bag. After her mother gave her $42, she was still short of 1/4 of the cost of the hand bag. How much did the hand bag cost?

Answer: ....

PROBLEM 193

Shelly has 60 dollars and she wants to buy some gifts for her mother. She first buys a box of chocolates. With half of the money she had left, she buys a heart necklace. She then spends one fourth of the remainder on a dozen roses. When Shelly returns home, she has 15 dollars left. How much money did the box of chocolates cost?

Answer: ....

PROBLEM 194

The total cost of 2 T-shirts and 3 skirts is $234. If 3 such T-shirts and 1 skirt cost $120, find the cost of each T-shirt.

Answer: ....

PROBLEM 195

Wendy saw a cosmetic shop having a 15% off sale on all their products. Wendy bought 9 lipsticks from the shop. Each lipstick cost $4.25 after the discount. How much money did she save?

Answer: ....

PROBLEM 196

A parking lot has a rate of 40 cents for the first two hours and 15 cents per hour for each succeeding. If the owner of a car paid $1.25 cents in all, how long did his car stay in the parking lot?

Answer: ....

PROBLEM 197

2 pens cost as much as 15 pencils. If 20 pens and 30 pencils cost $90, how much did each pencil cost?

Answer: …………………………

William and his sister had a total of $520. After he spent ½ of his money and his sister spent 1/5 of her money, both of them still had the same amount of money left.

a) How much money did he spend?

b) How much money did his sister have left?

Word Problem Multiplication
76A 2

Answer: ………………………

PROBLEM 199

My mom was 27 when I was born. 8 years ago she was twice as old as I shall be in 5 years time. How old am I now?

Answer: ....

PROBLEM 200

A book costs $16 more than a magazine. The total cost of 3 books and 2 magazine is $68. What is the total cost of 5 books and 5 magazines?

Answer: ....

DETAILED SOLUTIONS

6

Guaranteed to improve children's math and success at school.

1. Numbers

Solution to Question 1

Nth term = $(n+1)^2$
The 11th term will be $(11+1)^2 = 12^2 = 144$

Solution to Question 2

We have to find:
1 + 2 + 3 + ... + 119 + 120
A shortcut method to find the sum of n consecutive numbers is = $\dfrac{(n) \times (n+1)}{2}$
So 1 + 2 + 3 + ...+ 119 + 120

$= \dfrac{120 \times 121}{2}$ = 60 x 121 = 7260

Solution to Question 3

Following order of operations we solve x and ÷ first and then do + and −
12 + 20 ÷ 4 − 16 x 2 + 32 ÷ 2
= 12 + 5 − 32 + 16
= 17 − 32 + 16
= − 15 + 16
= 1

Solution to Question 4

The sum of n consecutive numbers from 1 to n is =

$\dfrac{n \times (n + 1)}{2}$

= (100 x 101)/2
= 5050

218

Solution to Question 5

Let the first number be X.

Then the five consecutive numbers will be X, X+1, X+2, X+3, X+4

X + X + 1 +X + 2 + X + 3 +X + 4 = 5920

5X + 10 = 5920

5X = 5920 − 10

X = 5910/5 = 1182

So the numbers are 1182, 1183, 1184, 1185, 1186

The sum of all the digits of these 5 numbers

= 1 + 1 + 8 + 2 + 1 + 1 + 8 + 3 + 1 + 1 + 8 + 4 + 1 + 1 + 8 + 5 + 1 + 1 + 8 + 6

= 70

Solution to Question 6

25 consecutive integers add up to 700.

Let the first number be X

Then the 25 numbers in the series will be:

X, X + 1, X + 2, X + 3, X + 4, X + 5, X + 6, X + 7, X + 8, X + 9, X + 10, X + 11, X + 12, X + 13, X + 14, X + 15, X + 16, X + 17, X + 18, X + 19, X + 20, X + 21, X + 22, X + 23 and X + 24

We are given their sum = 700

So 25X + 300 = 700

25X = 700 − 300

X = 400/25 = 16

The largest of these numbers = 16 + 24 = 40

219

Solution to Question 7

The sum of the squares of the first 20 positive integers is 2870.
Last five positive integers in this list are 20, 19, 18, 17 and 16
Sum of squares of last 5 integers = 400 + 361 + 324 + 289 + 256 = 1630
The sum of the squares of the first 15 positive integers
= 2870 − 1630 = 1240

There is also a shortcut formula for calculating the squares of first n numbers (starting from 1)

$$= \frac{(n) \times (n+1) \times (2n+1)}{6}$$

So when we are finding the sum of squares of the first 15 numbers, it is

$$= \frac{15 \times (15 + 1) \times (2 \times 15 + 1)}{6}$$

$$= \frac{15 \times 16 \times 31}{6}$$

$$= 1240$$

Solution to Question 8

Let the second number be X.
The first number is ten times the second number, so the first number is 10X
The sum of the two numbers is 5016
So:
X + 10X = 5016
11X = 5016
X = 5016/11 = 456
The numbers are 4560 and 456
Difference between the two number's = 4560 − 456 = 4104

220

Solution to Question 9

Let one number be X and then other number will be X + 8

Product = X x (X+8) = 48

We have to factorize 48 into two parts such that one of them is 8 more than the other.

The ways to factorize 48 are:

1 x 48

2 x 24

3 x 16

4 x 12

6 x 8

Of these 4 and 12 are such that one is 8 more than the other.

So the two numbers are 4 and 12.

Solution to Question 10

Let the 2 numbers be X & Y

Pete subtracted two numbers and the difference was 12.

X – Y = 12

X = 12 + Y

Michelle multiplied the same two numbers and the product was 540.

XY = 540

Substitute the values of X in the above equation

(12 + Y) x Y = 540

We have to factorize 540 into 2 parts so that one part is 12 more than the other.

540

=2 x 270	= 3 x 180	= 4 x 135	= 5 x 108
= 6 x 90	= 9 x 60	= 10 x 54	= 12 x 45
= 15 x 36	= 18 x 30	= 20 x 27	

So the numbers are 18 and 30

Sum = 18 + 30 = 48

Solution to Question 11

Let us write the factors of 72
The way to split 72 is:

1 x 72   sum = 73
2 x 36   sum = 38
4 x 18   sum = 22
6 x 12   sum = 18
8 x 9   sum = 17

The factors are both greater than the lowest even number, which is 2
So the only possible options are 4 x 18, 6 x 12 or 8 x 9.
The sum has to be a prime number so the numbers are 8 and 9

Solution to Question 12

The number of palindrome number's between 100 to 1000 are 90

From 100 to 200, the number of palindromes are:
101,111,121,131,141,151,161,171,181,191
These are 10 palindromes.
The same pattern will follow in subsequent set of 100 numbers also.
Number of palindromes from:

200 – 300 = 10
300 – 400 = 10
400 – 500 = 10
500 – 600 = 10
600 – 700 = 10
700 – 800 = 10
800 – 900 = 10
900 – 1000 = 10

Total number of palindromes from 100 to 1000 = 90

Solution to Question 13

The number of possible way to seat 4 people in

ABCD	BACD	CABD	DABC
ABDC	BADC	CADB	DACB
ACBD	BCAD	CBAD	DBCA
ACDB	BCDA	CBDA	DBAC
ADBC	BDAC	CDAB	DCAB
ADCB	BDCA	CDBA	DCBA

Solution to Question 14

1st clock chimes every 4 min
2nd clock chimes every 6 minutes
3rd clock chimed at 10.40 am and 10.48 am
Therefore the time gap between the 2 chimes = 8 min
The LCM of 4,6,8 is 24
So every 24 min they all will chime together.
Next time they will all chime together = 2:24 pm

Solution to Question 15

The sand in hourglass 1 runs out in 7 minutes
The sand in hourglass 2 runs out in 9 minutes.
We need to calculate the LCM of 7 and 9
LCM of 7 and 9 is 7 x 9 = 63 minutes
Peter will reset the hour glasses at the same time every 63 min i.e. 1 hr 3 min
If Peter reset the hourglasses together at 2.57pm, the next time he will reset both the hourglasses at the same time = 2:57 + 1:03 = 4pm

Solution to Question 16

The mean of the heights of five basketball players = 1.88 meters.
Sum of heights of 5 players = 1. 88 x 5 = 9.4 meters
When two new players with the heights of 1.81m and 1.93 meters are added
The new mean height of the basket ball players
= (9.4 + 1.81 + 1.93)/7 = 13.14/7 = 1.87 meters

Solution to Question 17

When tables are joined, only 2 students can be seated in the middle tables and
the first and last tables 3 students cab be seated
Let the number of middle tables be X
For X tables number of students students that can be seated = 2X
2X + 2 x 3 = 200
X = (200 − 6)/2
X = 194/2 = 97
Number of tables required = 97+2 = 99

Solution to Question 18

A + B = 120
C = 30
A + B + C + D = 170
D = 170 − (A+B) − C
   = 170 − 120 − 30 = 20 grams

Solution to Question 19

Let AB be the Eiffel tower and C where Carl is standing
It is given that BC = 300 and AC = 500
Using Pythagoras theorem,
$AB2 + BC^2 = AC^2$

224

$AB2 = AC^2 - BC^2$

$AB = \sqrt{AC^2 - BC^2}$

$\sqrt{500^2 - 300^2} = \sqrt{250000 - 90000}$

$= 400$

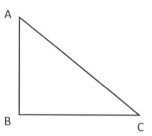

The Eiffel tower is 400 meters high.

Solution to Question 20

Average height = total height ÷ Number of players
a)Average = (1.72m+ 1.86m+ 1.84m+ 1.92m + 1.78m + 1.80)/6 = 1.82 meters
Total height of 6 players = 10.92 meters
b)4 years later, 3 of the players grew by 6cm = 3 x 6/100 = 0.18m
New total height = 10.92 + 0.18 = 11.1 meters
New average = (10.92 + 0.18)/6 = 1.85 meters

Solution to Question 21

The number of sweets has to be such that 1 less than it is a multiple of 3 and 5.
And it should be a multiple of 7.
The LCM of 3 and 5 is 15.
Multiples of 15 that are between 50 and 100 are:
60 75 and 90.
Of these numbers, if we add 1, only 90 becomes a multiple of 7.
So the number of sweets is 91.

Solution to Question 22

We can see that

1st figure is a 4 X 4 square and has a 2x2 squares colored.

2nd figure is a 5 X 5 square and has a 3x3 squares colored.

a) The nth figure will be a (n+3) x (n+3) square and will have a

(n+1) x (n+1) colored squares.

Number of uncolored squares = (n+3)2 − (n+1)2

In the 5th figure we will have a (5 +3) x (5 + 3) square and

(5 + 1) x (5 + 1) colored squares. So it will have 8 x 8 = 64 squares, number of

colored squares = 6x 6 = 36 and the number of uncolored squares will be 64 − 36

= 28

b)11th figure will have 14x 14 = − 196 squares, it will have 12 x 12 = 144 colored

squares and the number of uncolored squares =

196 − 144 = 52

Solution to Question 23

Let us first write equations for the cases mentioned above:

(A + B + C + D)/4 = 140kg   A + B + C + D = 560

(D + E + F)/3 = 140 − 20 = 120kg   D + E + F = 360

A = E + 20kg   A   = E + 20

(B + C + D) /3 = 160kg   B + C + D   = 480

F = 120kg   F   = 120

A + B + C + D = 560   and B + C + D = 480

So A + 480 = 560 A = 560 − 480 =   A = 80 kilograms

A = E + 20   80 = E + 20   E = 80 − 20   E = 60 kilograms

D + E + F = 360   D + 60 + 120 = 360   D = 180 kilograms

So weight of D is 180 kilograms

To find the overall average, we need to do (A + B + C + D + E + F) ÷ 7

A + B + C + D = 560 and

D + E + F = 360

Add both the equations and subtract the value of D to get (A + B + C + D + E + F)

(A + B + C + D + E + F) = 560 + 360 − 180 = 740 Kg.

Average weight of all boxes = 740 ÷ 7 = 105.7 Kg

226

Solution to Question 24

Jim, Joe and Julie will eat together again when the time passed is the LCM of 6, 7.5 and 12.

The LCM of 6, 7.5and 12 is 60.

So after 60 minutes they will eat a nut together.

The time when they eat a nut together is 2.30 + 1.00 = 3.30 PM

Solution to Question 25

No of juice cans used to form this arrangement =

9 + 8 + 7 + 6 + 5 + 4 + 3 + 2 + 1

= 45

Solution to Question 26

Athlete A runs 6 full circles in 120 minutes.

Time taken for 1 circle = 120/6 = 20 min

Athlete B makes 5 full circles in 60 minutes.

Time taken for 1 circle = 60/5 = 12 min

They will both be at the starting point at the time that is a multiple of both 20 and 12.

The LCM of 20 and 12 is 60.

If they start running now from the same point, they will be at the same starting point again after 60 min.

Solution to Question 27

If there are two people at a party, they can shake hands once. There is no one else left to shake hands with.

i.e. 2 people, 1 handshake

If there are three people at a party, the first person can shake hands with the two other people (two handshakes). Person two has already shaken hands with person one, but he can still shake hands with person three (one handshake). Person three has shaken hands with both of them.

i.e. 2 + 1 = 3.

3 people = 3 handshakes

If there are four people at a party, person one can shake hands with three people, person two can shake hands with two new people, and person three can shake hands with one person.

i.e. 3 + 2 + 1 = 6.

4 people = 6 handshakes

Extending the same the pattern, the number of handshakes with 9 people will be:

8 + 7 + 6 + 5 + 4 + 3 + 2 + 1 = 36

Solution to Question 28

Average marks for 3 tests = 56

Total marks in 3 tests = 56 x 3 = 168

Let the number of marks to be scored in 4th test be X

$(168 + X)/4 = 65$

$168 + X = 65 \times 4$

$X = 260 - 168 = 92$

Kathleen should score 92 marks to get an overall average of 65.

Solution to Question 29

$$\frac{x - xy}{(1 - y)(1 + y)} = \frac{x(1 \not/ y)}{(1 \not/ y)(1 + y)} = \frac{8}{(1 + y)}$$

$$\frac{8 - 48}{(-5)(7)} = \frac{8 - 6 \times 8}{(1 - 6)(1 + 6)} = \frac{8}{1 + 6} = \frac{8}{7}$$

Factors of 72 = 2 x 2 x 2 x 3 x 3
Factors of 126 = 2 x 3 x 3 x 7
Therefore HCF of 72,126 = 2 x 3 x 3 = 18
HCF of 72 and 126 is 18
Number of girls teams = 72/18 = 4
Number of boys teams = 126/18 = 7
Total number of teams = 4 + 7 = 11

Let the number be a, b, c, d, e, f
The average of six numbers in a list is 64.

$$\frac{a + b + c + d + e + f}{6} = 64 \quad (1)$$

The average of the first two numbers is 26.

$$\frac{a + b}{2} = 26$$

a + b = 26 x 2 = 52
Substitute the value of a+b in (1)
52 + c + d +e +f = 64 x 6
c + d + e + f = 384 − 52 = 332
The average of the last four numbers = 332/4 = 83

Solution to Question 32

Let the letters delivered in 4th hour be X

$$\frac{140 + 120 + 170 + X}{4} = 150$$

$430 + X = 150 \times 4$

$X = 600 - 430$

$X = 170$ Letters to be delivered in 4th hour = 170

Solution to Question 33

18 children have a brother.

Let the number of children who have a brother and a sister be X.

Since 12 children have a sister, this includes the children who have a sister and a brother.

So the number of children who have just a sister = $12 - X$

Similarly, the number of children who have a brother is 18. So the number of children who have only a brother = $18 - X$ (Refer figure)

Since the total number of children is 29, we have:

$12 - X + X + 18 - X + 3 = 29$

$33 - X = 29$

$X = 4$

So 4 children have a sister and a brother.

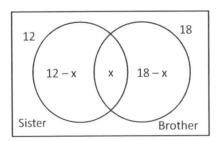

2 Algebra

Solution to Question 34

Let the total number of books = X
Chinese books = 1/4 X
Remaining books = 3/4 X
English books = 3/4 x 1/5 x X = 3X/20
Spanish books =3/4 x 4/5 x X = 12X/20
There are 63 more Spanish books than English books
3X/20 + 63 = 12X/20
9X/20 =63
X = 63 x 20/9 = 140
Number of Chinese books= 1/4 x 140 = 35
Number of English books = 3/20 x 140 = 21
Number of Spanish books = 12/20 x 140 = 84

Solution to Question 35

Let John have X marbles at first
Jane gave some marbles, then john had 3 times as many marbles as he had = 3X
After giving John, Jane had half of the number of marbles john had in the
beginning = 3X/2
Total number of marbles = 180
Therefore 3X/2 + 3X = 180
X = 180/4.5 = 40
Marbles with Jane at the beginning = 180 − 40 = 140

Solution to Question 36

Let X be the number of correct answers.

The number of wrong answers would be 15 − X

Each correct answer is awarded 3 points and 1 point is deducted for every wrong answer deducts 1.

So 3 x X − (15 − X) x 1 = 29

3X − 15 + X = 29

4X = 29 +15 = 44

X = 11

Number. of wrong answers = 15 − 7 = 8

Solution to Question 37

Let the January salary of Mrs Jones be X.

Savings in January by Mrs Jones = 10X/100 = X/10

Saving in February = X/10 + (50/100 x X/10)

                 = X/10 + X/20 = 3X/20

X/10 + 800 = 3X/20

3X/20 − X/10 = 800

3X − 2X = 800 x 20

X = $16000

Monthly salary of Mrs Jones = $16000

Solution to Question 38

Total number of students in the class = 42

Number of students who passed in the science examination = 2/3 x 42 = 28

Number of students who failed in science examination = 42 − 28 = 14

Let the number of boys who failed in science be 'B'

The Girls who failed in science examination = B + 4

B + B + 4 = 14

2B = 14 − 4

B = 10/2 = 5

Girls who failed = 5 + 4 = 9
Let the girls who passed the science examination be 'G'
Therefore boys who passed the science examination = G + 80G/100 = 180G/100
G + 180G/100 = 28
100G + 180G = 2800
280G = 2800
G = 2800/280 = 10 i.e Girls who passed in the science examination = 10
Boys who passed in the science examination = 10 x 180/100 = 18
Total number of girls in the class = 9 + 10 = 19
Total number of boys in the class = 5 + 18 = 23

 **Solution to Question 39**

Let the number of people in the party be X.
1/3 of the people are men = X/3
Women = 2X/3
Half of the men left the party = 1/2 x X/3 = X/6
There were 192 more women than men
X/6 + 192 = 2X/3
2X/3 – X/6 = 192
4X – X = 192 x 6
3X = 1152
X = 1152/3 = 384
Men = 384/3 = 128
Men left in the party = 128/2 = 64
Women in the party= 384 x 2/3 = 256
16 women left the party. The remaining women in the party = 256 – 16 =240
Ratio of the number of women to the number of men left at the party
= 240 : 64
= 15 : 4

Solution to Question **40**

Let X be the number of days over which Freddy and Justin spend the money.

16X + $70 = 8X + $430

8x = 430 − 70 = 360

 X = 45

8 x 45 + 430 = $360 + $430 = $790

Their parents gave them $790 each at first.

Solution to Question **41**

Mass of empty Tub A = Mass of empty Tub B = Mass of empty Tub C = X

When a tub is full, let the weight of water be W

X + 0.5 W = 3.1 kilograms

X + 0.2 W = 2.2 kilograms

Subtracting

0.3W = 0.9 kilograms

    W = 3 kilograms

Substituting into the first equation,

X + 1.5 kilograms = 3.1 kilograms

X = 1.6 kilograms

Now, assume Y to be the fraction of the water capacity of Tub C when it weighs 2.8 kilograms

1.6kg + Y x 3 = 2.8 kilograms

3Y = 1.2 kilograms

Y = 0.4 = 2/5

Tub C must be 2/5 full to have a mass of 2.8 kilograms.

234

Solution to Question 42

Let A be the cost of an apple and O be the cost of an orange

2 A + O = 1.20

2 O + A = 1.80

Adding the two equations we get

3A + 3O = 3.00

A + O = 1.00

2A + O = 1.2 and

A + O = 1.00

So A = 0.2 and O = 0.8

The cost of an apple = $0.2 and the cost of an orange = $0.8

Solution to Question 43

Let the number of toys with Adrian be X and with Judy be Y.

X + 36 = Y

X = Y − 36

They were then given an equal number of toy cars and Judy had now twice the number of toy cars she had initially.

That means number of cars given = Y

Judy now has = 2Y cars and Adrian has Y − 36 + Y cars

1/3 of Judy's cars were now 1/2 of Adrian's cars.

1/3 x 2Y = 1/2 x (Y − 36 + Y)

2Y/3 = 1/2 x (2Y − 36)

4Y = 6Y − 108

2Y = 108

Y = 108/2 = 54

X = 54 − 36 = 18 cars

a) Number of Toy cars given to each of them = 54

b) Number of Toy cars Adrian had at first = 18

Solution to Question 44

Let the number of students in a class be X.
20% are boys = 20X/100
Number of Girls = 80X/100
20% of the boys keep pets= 20X/100 x 20/100
10% of the girls keep pets = 10/100 x 80X/100
144 girls do not keep pets
144 = 90/100 x 80X/100
144 = 72X/100
X = 144 x 100/72 = 200
Number of students in the class = 200

Solution to Question 45

Let the number of boys be X.
Number of girls = X + 6
Each Girl paid $10 and each boy paid $1 less = $9
(X+6) x 10 + X x 9 = 364
10X + 60 + 9X = 364
19X = 364 − 60
X = 304/19 = 16
Number of boys in the group = 16

Solution to Question 46

Let the number of kittens be X.
There are four times as many kittens as canaries in the shop.
Canaries = 4X
Let the number of puppies be Y
When the shop assistant makes a count of the animals, she find that there are 30 heads and 88 legs.
X + 4X + Y = 30 (heads)
4X + 8X + 4Y = 88 (legs: canaries have 2 legs, kittens 4 legs, puppies 4 legs)
5X + Y = 30

236

Y = 30 – 5X
Substitute this value in the equation below
12X + 4Y = 88
12X + 4(30 – 5X) = 88
12X + 120 – 20X = 88
8X = 120 – 88
X = 32/8 = 4
a) Number of Canaries in the shop = 16
b) Number of Puppies in the shop = 10

Solution to Question 47

Let the number of cookies be 'B' and chocolate chips be 'C'
B + C = 120
B = 120 – C
2B + 4C = 310
2 x (120 – C) + 4C = 310
240 – 2C + 4C = 310
240 +2C = 310
2C = 310 – 240
C = 70/2 = 35
B = 120 – 35 = 85
Number of butter cookies sold = 85
Number of chocolate chip cookies sold = 35

Solution to Question 48

Let the number of roosters be X
A farmer had twice as many chicken as roosters , so the number of chicken = 2X
After he sold 244 chicken, he had half as many chicken as roosters.
2X – 244 = X/2
2X – X/2 = 244
4X – X = 244 x 2
3X = 488
X = 488/3 = 163
No of roosters = 163

Solution to Question 49

Let the number of oranges sold be X.
Number of Pears sold = X + 10X/100 = 1.1 X
Number of Apples sold = (1 + 20/100) x (1.1X) = 1.32 X
1.32 X = 198
X = 198/1.32 = 150
 Number of oranges sold = 150
Number of pears sold = 165
Number of apples sold = 198
Number of packets sold:
Apples = 198 / 6 =33
Pears = 165 / 11 = 15
Oranges = 150/10 = 15
Total money earned = 33 x 2 + 15 x 3 + 3 x 15 = 66 + 45 + 45
= $156

Solution to Question 50

Let the number of beads with Mark's sister be X.
Number of beads with Mark = X + 4100
After Mark gave 900 beads to his sister, he had thrice as many beads as his sister.
X + 4100 – 900 = 3(X + 900)
X + 3200 = 3X + 2700
2X = 3200 – 2700
X = 500/2 = 250
Number of beads Mark had at first = 250 + 4100 = 4350 beads

Solution to Question 51

Let Colin's salary be $X
He gives 20% of his salary to his wife = 20/100 x X = X/5
He spent 30% of the remainder = 30/100 x 4X/5 = 12X/50
Money left after that = 70/100 x 4X/5 = 28X/50
He gave the rest of the money to his sons in the ratio 5 : 3

Jason get = 5/8 x 28X/50 = 7X/20

James gets = 3/8 x 28X/50

The % of his salary that Jason gets from Colin

= 7X/20 x 100/X = 35%

 ## Solution to Question 52

Let the Chinese books be X

Number of English books = X + 1647

When 2400 English books were sold out, there were 4 times as many Chinese books than English books left in the book store.

(X + 1647 − 2400) x 4 = X

4X − 753 x 4 = X

3X = 3012

X = 3012/3 = 1004

Total number of books in the store at first = 1004 + 1004 + 1647 = 3655

 ## Solution to Question 53

Let Janice's savings in the bank be $X.

She spent 25% of her savings on clothes and the remaining amount on a television set and a hi-fi set in the ratio 2:3.

Money spent on Clothes = 25X/100

Remaining amount = X − 25X/100 = 75X/100

Amount of money spent on Television = 2/5 x 75X/100

Amount of money spent on Hi-fi set = 3/5 x 75X/100

She spent $280 more on the Wi-Fi set than on clothes.

25X/100 + 280 = 3/5 x 75X/100

3/5 x 75X/100 − 25X/100 = 280

45X/100 − 25X/100 = 280

20X/100 = 280

X = 280 x 100/20 = $1400

Her original savings = $1400

Solution to Question 54

Let the amount that Janice paid be $X.

Fred paid = 2X

Marcus paid = 3 x 2X = 6X

X + 25 = 6X

5X = 25

X = 5

Cost of dinner = 5 + 10 + 30 = $45

Solution to Question 55

Let the amount of money with Terry be $T and with Sarah be $S

T + S = 745

S = 745 − T

Mother gave Sarah $15 and then Terry had 3 times as much money as her

S = 745 − T + 15 = 760 − T

3 (760 − T) = T

2280 − 3T = T

4T = 2280

T = 2280/4 = $570

S = 745 − 570 = $175

Amount of money that Terry had more than Sarah = 570 − 175 = $395

Solution to Question 56

Let the amount of savings that Thomas has be $X

Sam's savings are 10% more than Thomas.

Sam's savings = X + 10X/100 = 110X/100

After Sam transfers $240 to Thomas's account, they have the same amount of savings

110X/100 − 240 = X + 240

110X − 24000 = 100X + 24000

10X = 24000 + 24000

X = 48000/10 = $4800

Sam's saving = 110/100 x 4800 = $5280

240

Solution to Question 57

Let the number of Pupils in the hall at first be X.

Number of Girls = 80X/100 = 4X/5

The remaining number are boys = X/5

After 140 people entering the hall,

number of girls increases by 25%.

Number of new girls = 4X/5 + (4X/5 x 25/100) = 4X/5 + X/5 = X

Number of boys increases by 75%.

Number of new boys = X/5 + (X/5 x 75/100)= X/5 + 3X/20 = 7X/20

The new number of girls plus the new number of boys = initial number of Pupil + 140

X + 7X/20 = 140 + X

7X/20 = 140

X = 140 x 2 0 /7

X = 400

The number of Pupils in the hall now = 400 + 140 = 540

Solution to Question 58

Let the weight of flour be X kilograms.

Weight of rice = X + 160

After selling 400 kilograms rice and 30 kilograms flour, he has thrice as much flour than rice.

3 x (X + 160 − 400) = (X − 30)

3 x (X − 240) = X − 30

3X − 720 = X − 30

2x = 690

X = 345

Total weight of flour at first = 345 kilograms

Weight of rice = 345 + 160 = 505 kilograms

 **Solution to Question 59**

Let the number of right questions be X

The number of wrong questions = 30 − X

For every question he got right, he earned 10 points, and for every question he got wrong, he lost 2 points and he got a score of 120. So:

10X − 2 (30 − X) = 120

10X − 60 + 2X = 120

12X = 120 + 60

X = 180/12

X = 15

No of questions Gavin got right = 15

 **Solution to Question 60**

Let the number of 20 cents coins be X.

Number of 50 cents coins = X + 8

0.2X + (X + 8) x 0.5 = 60

0.7X + 4 = 60

0.7X = 56

X = 80 coins

Number of 20 cent coins = 80

Number of 50 cent coins = 80 + 8 = 88

Total number of coins in his piggy bank = 80 +88 = 168

 **Solution to Question 61**

Let the wallet's cost be X.

Cost of Pen = X + 7

Cost of Tie = X − 5

X + 7 + X − 5 = 21

2X +2 = 21

2X = 19

X = 19/2 = 9.5

Cost of Wallet = $9.5

242

Solution to Question 62

Let the amount of money with Betty be X
Amount of money with Alex = X + 1.50
Alex has 3 times as much money as Colin.
So money with Colin = (X+1.50)/3
The sum of their money = $11.8
X + 1.5 + X + (X + 1.5)/3 = 11.8
6X + 4.5 + X + 1.5 = 11.8 x 3
7X + 6 = 35.4
7X = 35.4 − 6 = 29.4
X = 29.4/7 = $4.2
The amount of money with Colin = $(4.2 + 1.5)/3 = $1.9

Solution to Question 63

Let the age of Allan, Bernard and Cindy be A, B and C respectively

Then   A + B = 20
        B + C = 21
        C + A = 25

If you add up the three equation, you get

        A + B + B + C + C + A = 20 + 21 + 25
        2(A + B + C) = 66
        A + B + C = 33
        Since B + C = 21
        A + 21 = 33
        A = 33 - 21 = 12 years

Allan is 12 year old.

D = 88/8 = 11 years
A = 72 − 3D = 72 − 33 = 39 years
Putting these back in the equations above we get B + C = 25 years.
We don't have any more information to find the ages of Bernard and Cindy

 ## Solution to Question 64

There were 9 ten-cent coins and the rest were twenty-cent coins and fifty-cent coins.
Number of twenty-cent and fifty-cent coins = 46 − 9 = 37
Let the number of 20 cents coins be X and the number of 50 cent coins be 37 − X
Since the total value of the money is $16.10 (or 1610 cents), we have
9 x 10 + 20X +50 x (37 − X) = 1610
90 + 20X + 1850 − 50X = 1610
1940 − 30X = 1610
30X = 1940 − 1610
X = 330/30 = 11
Number of 20-cent coins = 11
Number of 50-cent coins = 37 − 11 = 26

 ## Solution to Question 65

Let the age of Joe be X
Lucy is 10 years older than Joe = X + 10
Mark is 7 years older than Lucy = X + 10 + 7 = X + 17
The sum of their ages is 45 years.
X + X + 10 + X + 17 = 45
3X + 27 = 45
3X = 45 − 27
X = 18/3 = 6 yrs
Joe = 6 yrs   Lucy = 6 + 10 = 16 yrs   Mark = 6 + 17 = 23 yrs

Solution to Question 66

A school has 10 classes with the same number of students in each class.

Let the number of students in a class be X.

One day, many students were absent.

5 classes were half full = X/2 * 5 = 5X/2

Number of students present = 5X − 5X/2 = 5X/2

3 classes were 3/4 full = 3 * 3X/4 = 9X/4

Number of students present = 3X − 9X/4 = 3X/4

2 classes were 1/8 empty = X/8

A total of 50 students were absent.

5X/2 + 3X/4 + X/8 = 50

(20X + 6X + X)/8 = 50

27X = 50 * 8

X = 400/27

X = 14.8 = 15

No of students in this school when no students are absent = 15 * 10 = 150

Solution to Question 67

The apples and oranges weigh 15 kg together.

A + O = 15

The oranges and peaches weigh 18 kg together.

O + P = 18

The apples and peaches weigh 16 kg together.

A + P = 16

Add all the equations

A + O + O + P + A + P = 15 + 18 + 16

2 * (A + O + P) = 49

A + O + P = 49/2 = 24.5 kg

No of kgs of fruits Samantha bought altogether = 24.5 kg

Solution to Question 68

Let the length of the second part of wire be X
Length of the longest piece of wire= 2X
The length of the third piece of wire= 21 cms
X + 2X + 21 = 120
3X = 120 − 21
X = 99/3 = 33 cm
The length of the longest piece cut form the wire = 33 x 2 = 66 cm

Solution to Question 69

Let the no of students in school C be X
The number of students in school B is 60% that of school C = 60X/100
Total number of students in school A is 30% that of school B = 30/100 x 60X/100 = 18/100 X
1600 students were transferred from school C to school A, the number of students in school C will be 1/6 of the number of students in school B.
X − 1800 = 1/6 x 60X/100
X − 1800 = 10X/100 = X/10
X − X/10 = 1800
9X/10 = 1800
X = 1800 x 10/9
X = 2000
Number of students in school C = 2000
Number of students in school B = 60/100 *2000 = 1200
Number of students in school A = 30/100 x 1200 = 360
Total number of students in the three schools = 2000 + 1200 + 360 = 3560

Solution to Question 70

Let the amount of water poured be X
4.2 + X = 5 x(0.6 + X)
4.2 + X = 3 +5 X
5X − X = 4.2 − 3
4X = 1.2
X = 1.2/4 = 0.3 L
a) Amount of water poured into each container = 0.3 L
b) The total amount of water in the 2 containers now =
0.6 + 0.3 + 4.2 + 0.3 = 5.4 L

246

Solution to Question 71

Let the initial length of the ribbon be X

Emily gave one third of it to her sister = X/3

Remaining ribbon with her = X – X/3 = 2X/3

She used 40% of the remaining on her dress = 40/100 x 2X/3

She was left with 60/100 x 2X/3 = 40X/100

She cut this into 4 equal parts and gave one to each of her cousins.

So each cousin got: 40X/100 x ¼ = 10X/100 cms.

Since each cousin got 15 cms,

So 10X/100 = 15

X = 15 x 10 = 150 cms

The length of the ribbon used on the dress = 40/100 x 2/3 x 150 = 40 cm

Solution to Question 72

Let the weight of the jar be J

Let the mass of each nail be N

And the weight of each screw be 2N

20N + J = 0.5kg or J = 0.5 – 20N

40x2N + J = 1.4kg

80N + J = 1.4

Putting the value of J from above we get:

80N + 0.5 – 20N = 1.4

60N = 1.4 – 0.5

60N = 0.9 kilograms = 900 grams

N = 900/60 = 15 grams

J = 0.5 – 20N

J = 500 – 20 x 15

J = 500 – 300 = 200 grams

Solution to Question 73

Let the number of cows be X.

Therefore the number of chickens = 45 − X

Cow has 4 legs where as chickens have 2 legs.

4X + 2 (45 − X) = 126

4X + 90 − 2X = 126

2X = 126 − 90

X = 36/2 = 18

Number of cows on Justin's farm = 18

Solution to Question 74

Let the number of correct answers be X

Number of wrong answers = 25 − X

For every question he got right, Richard earned 10 points, and for every question he got wrong, he lost 2.5 points.

He answered every question, and got a score of 150.

10X − 2.5 x (25 − X) = 150

10X − 62.5 + 2.5X = 150

12.5X = 150 + 62.5

X = 212.5/12 = 17

Number of questions Richard got right = 17

Solution to Question 75

Total number of men = 105

Men who left the party = 2/7 x 105 = 30

Men remaining in the party = 5/7 x 105 = 75

Let the total number of persons in the party be X.

40% of adults left the party = 40 /100 x X

10% of those are men = 10/100 x 40/100 x X = 30

X = 30 x 100/10 x 100/40 = 750

Therefore number of women in the beginning = 750 − 105 = 645

Solution to Question 76

Let B be the number of Bananas in the beginning.
NUmber of Bananas sold = 7
Number of Bananas sold is 1/7 of original number of Bananas = 1/7 x B = 7
There were 49 Banana's in the beginning
Number of Bananas left = 49 – 7 = 42
42 = (number of Oranges left) /4 (as number of bananas left was 1/4 of the number of Oranges left)
Number of Oranges left = 42 x 4 = 168
As 1/3 of the Oranges were sold, number of Oranges in the beginning = 168 x 3 = 504
Therefore number of Bananas in the beginning = 49,
Number of Oranges in the beginning= 504

3. Fractions , Decimal and Percentages

Solution to Question 77

The number of questions marked correctly = 85/100 x 45 = 38.25
Marks scored in the second test = 38.25 x 4 = 153
In the previous week's test, number of questions marked correctly = 65/100 x 45 = 29.25
Marks scored in the same test previous week = 29.25 * 4 = 119
Number of questions Charles answered correctly on this week's test more than the last week's test = 38.25 – 29.25 = 9

Solution to Question 78

Let the total number of beads be X.
Number of green beads = 40% of X = 40X/100
Number of yellow beads = 60% of X = 60X/100
He lost 50 beads and the number of beads fell to 2/3 of the initial beads
60X/100 – 50 = 2/3 x 60X/100
1/3 x 60X/100 = 50
X/50 = 50
X = 50 x 50 = 250
In the end Tommy had = 250 – 50 = 200 beads

Solution to Question 79

Let the total number of coins be X.

It is given that 2/3 coins are 20 cents, so the remaining coins are of 5 cents. So 1/3 of the coins are 5 cent coins

2/3 x X x 0.2 + 1/3 x X x 0.05 = 7.20

0.4X + 0.05X = 7.20 x 3

0.45X = 21.6

X = 21.6/0.45 = 48

Number of 20 cents coins = 2/3 x 48 = 32

Number of 5 cents coins = 48 − 32= 16

Solution to Question 80

Let the number be X.

2/3 x X = 1/2 x X + 12

4X = 3X + 72

X = 72

Third multiple of the number = 72 x 3 = 216

The difference of the number and its 3rd multiple = 216 − 72 = 144

Solution to Question 81

There were a total of 2250 seats in a theatre.

10% of the seats were first class = 10/100 x 2250 = 225

30% of the seats were second class = 30/100 x 2250 = 675

and the rest of the seats were third class = 2250 − 225 − 675 = 1350

2 years later, 100 first class seats and 125 second class seats were added.

First class seats in the end = 225 + 100 = 325

Second class seats in the end = 675 + 125 = 800

The total number of seats in the theatre = 2250 + 125 + 100 = 2475

Percentage of seats that were third class in the end = 1350/2475 x 100 = 54%

Solution to Question 82

Let the number of men in the contest be X.

There were thrice as many men as women.

Women = 3X

3/5 of the female participants were children = 3/5 x 3X = 9X/5

1/4 of the male participants were children = 1/4 x X = X/4

X + 3X + 9X/5 + X/4 = 847

20X + 60X + 36X + 5X = 847 x 20

121X = = 847 x 20

X = 7 x 20

X = 140

Men = 140

Women = 3 x 140 = 420

Total number of adults = 420 + 140 = 560

Solution to Question 83

It is given that total number of cards Charles and Ryan had = 432

Let Charles have X cards and Ryan have 432 − X cards.

Charles gave 2/5 of his cards = 2X/5.

Remaining number of cards with Charles = X − 2X/5 = (5X − 2X)/5 = 3X / 5

Number of Cards with Ryan = 432 − X + 2X/5 = 288 − 4X/5

Ryan gave 1/4 of his cards to Charles = (288 − 4X/5)/4

NUmber of Remaining cards with Ryan = ¾ x (288 − 4X/5)

Cards with Charles = 3X/5 + (288 − 4X/5)/4

Both of them had equal number of cards in the end

3(288 − 4X/5)/4 = 3X/5 + (288 − 4X/5)/4

216 − 3X/5 = 3X/5 + 72 − X/5

216 − 72 = 3X/5 + 3X/5 − X/5

144 = 6X/5 − X/5 = X

X = 144

Therefore Charles had 144 and Ryan had 432 − 144 = 288 cards at first

Let the total number of audience be X.

Number of Adults = X/6, Number of Children = 5X/6

60% of children are girls = 5X/6 x 60 /100 = X/2

There were 48 more girls than adults.

X/6 + 48 = X/2

X/2 – X/6 = 48

3X – X = 48 x 6

2X = 48 x 6

X = 48 x 6/2 = 48 x 3 = 144

a) Number of Girls = 5 x 144/6 x 60 /100 = 72

Number of Adults = 24

Number of Boys = 40/100 x 5X/6 = 48

Let the number of boys who left be Y.

(144 – Y) x 25 /100 = 48 –Y

144 – Y = 4 x (48 – Y)

144 – Y = 192 – 4Y

3Y = 192 – 144 = 48

Y = 16

b) So 16 boys left the concert half way.

Let the number of files be X.

Number of red files = 2X/3

Number of blue files = X/3

3/4 of red files were given away = 2X/3 x 3/4

Remaining red files = 1/4 x 2X/3

1/4 of blue files were given away = X/3 x 1/4

Remaining blue files = 3/4 x X/3

2X/12 + 3X/12 = 100

5X = 100 x 12

X = 1200/5 = 240

Number of files Carmen had at first = 240

Solution to Question 86

Harry gave 0.25 of the cards to his niece. He gave 0.25 x 60 = 15 cards
Number of Remaining cards = 60 – 15 = 45
3/5 of remaining cards were given to his cousin =
3/5 x 45 = 27 cards

Harry was left with 60 – 15 – 27 = 18 cards

Solution to Question 87

Let the amount of money with Rachel be X.
Rachel spent 1/5 of her money on skirt = X/5
Remaining amount of money = 4X/5
She spent 1/3 of the remaining money on 4 blouses and a belt = 1/3 x 4X/5
Rest of her money was spent on perfume = 2/3 x 4X/5
The belt cost 1/4 as much as the bottle of perfume = ¼ x (2/3 x 4X/5)
Money spent on each blouse = $21.5
Money spent for 4 blouses = 4 x 21.5 = $86
The amount of money spent on 4 blouses and a belt = 1/3 x 4X/5 = $86 + ¼ of
the cost of perfume
1/3 x 4X/5 = $86 + (1/4 x 2/3 x 4X/5)
4X/15 = 86 + 8X/60
4X/15 = 86 + 2X/15
4X/15 – 2X/15 = 86
2X/15 = 86
X = 86 x 15 / 2 = $645
Cost of perfume = 2/3 x 4/5 x 645 = $344

Solution to Question 88

a) Ratio of water to orange cordial= 25 : 1 5 = 5 : 3

% of Orange cordial in punch = 3 / 8 x 100 = 37.5%

b) Let's say she has to add X liters of orange cordial to make 50% in the mixture.

(15 + X)/ (40 + X) = 0.5

15 + X = 40 x 0.5 + 0.5X

X − 0.5X = 20 − 15 = 5

0.5 X = 5

X = 10 liters

Jenny should used 10 liters of orange cordial to 50% of orange cordial in the mixture.

Solution to Question 89

Length of wire = 16/5 = 3.2m

Length of one piece of wire = 3.2/8 = 0.4m

Total length of wire for making lantern = 5 x 0.4 = 2 meters

Solution to Question 90

Let Tom's weight be X.

Harry's weight = X + 37.2

Kathy's weight = 5/7 x (X + 37.2)

It is given that Tom's weight is 3/5 of Kathy's weight

3/5 x 5/7 x (X + 37.2) = X

3X + 111.6 = 7X

4X = 111.6

X = 111.6/4 = 27.9 kilograms

254

Solution to Question 91

Let the number of people at the Paris World Expo be X.
Number of Chinese = 2X/5
Number of French = 2X/5 + 32
Number of Dutch = 58
2X/5 + 2X/5 + 32 + 58 = X
4X/5 + 90 = X
X/5 = 90
X = 90 x 5 = 450

There were 450 people at the Paris World Expo.

Solution to Question 92

a) There were 120 mangoes in the basket.
2/3 of the mangoes were overripe = 2/3 x 120 = 80
Remaining were ripe = 1/3 x 120 = 40
40 mangoes were removed from the basket.
Remaining number of mangoes in the basket = 80
3/4 of the remaining mangoes were overripe = 3/4 x 80 = 60. That means 20 overripe mangoes were taken out
b) Since 40 mangoes were taken out of the basket, the number of ripe mangoes taken out of the basket = 40 − 20 = 20
Mangoes remaining in the basket are: 20 ripe and 60 overripe mangoes
c) There were 40 more overripe mangoes left in the basket than the ripe ones.

Solution to Question 93

600 pupils were asked to choose their favorite destination.
26% choose Thailand = 26/100 x 600 = 156
20% choose London = 20/100 x 600 = 120
12% choose Italy = 12/100 x 600 = 72
% of pupils choosing Hong Kong = 100 − 26 − 20 − 12 = 42%
Hong Kong = 42/100 x 600 = 252 pupils   252 - 120 =132
Number of more pupil who choose Hong Kong over London = 250 − 72 = 180

Solution to Question 94

Kelvin had 1200 apples.
5% of them were rotten and thrown away = 5/100 x 1200 = 60 apples
Remaining apples = 1200 – 60 = 1140
40% of the remainder were packed into boxes of 12 apples each and the rest into large boxes of 36 apples each.
Boxes of 12 will have = 40/100 x 1140 = 456 apples
Number of boxes of 12 = 456/12 = 38 boxes
Boxes of 36 will have = 1140 – 456 = 684 apples
Number of boxes = 684/36 = 19
The small boxes of apples were sold at $4.30 per box and the large boxes were sold at $13 per box
Total money collected   = 38 x 4.3 + 19 x 13
                          = 163.4 + 247
                          = $410.4

Solution to Question 95

Let the number of people in the squash club be X.
After 480 females left, the membership was decreased to 80% of its original enrolment.
X – 480 = 80/100 X
X – 480 = 0.8X
0.2X = 480
X = 480/0.2 = 2400
There were 2400 members in the club initially.

Solution to Question 96

Let the number of apples with the fruit seller be X.
1/5 is sold to Lena = X/5 =20X/100
35% of them to Caroline = 35X/100
Number of apples sold = 55X/100
Remaining apples = 45X/100

10 of them were rotten and he threw them away.

Net remaining apples left = 45X/100 – 10

Number of apples sold was 30 more than the number of apples left

55X/100 = 45X/100 – 10 + 30

55X/100 – 45X/100 = 20

10X/100 = 20

X = 20 x 100/10 = 200

Number of apples at the beginning = 200

Solution to Question 97

30% of the pupils in a lecture theatre were girls = 30/100 x 200 = 60

When some girls left the theatre, the percentage of girls dropped to 20%

If X girls left the theatre, the ratio of girls is 20%

(60 – X) / (200 – X) = 20/100

(60 – X) x 5 = (200 – X)

300 – 5X = 200 – X

300 – 200 = 5X – X

100 = 4X

X = 25

25 girls left the theatre.

Solution to Question 98

Let the total number of tarts with James be X.

25% of them are sold on Monday = 25X/100

80 fewer tarts are sold on Tuesday than on Monday

So number of tarts sold on Tuesday = 25X/100 – 80

On Wednesday James sold 160 tarts and found that he had 30% of his tarts left.

So 70% of tarts were sold

25X/100 + 25X/100 – 80 +160 = 70X/100

50X/100 +80 = 70X/100

20X/100 = 80

X = 80 x 100/20 = 400

Total number of tarts he sold = 70/100 x 400 = 280

Solution to Question 99

Let the amount of money with Rachael be X.

Spent half the money on wardrobe = X/2, $360 on a dressing table and had 1/8 left.

Remaining amount money X/2 = 360 + X/2 x X/8

X/2 – X/16 = 360

7X = 360 x 16

X = 5760/7 = $822

Solution to Question 100

There are some marbles in a bag. Bob takes 1/5 of them for himself.

After 1/5 of the marbles are taken, 1 – 1/5 = 4/5 marbles remain.

If John takes ¾ of the remaining marbles, then 1 – 3/4 = ¼ of the marbles remain.

¼ of 4/5 = ¼ x 4/5 = 1/5

So Tom gets 1/5 of the number of marbles. But he gets 4 marbles.

If 4 marbles are 1/5 of the marbles, then the number of marbles in the beginning = 4 x 5 = 20

There were 20 marbles in the beginning.

Solution to Question 101

Total number of students = 40

It is given that 40% of karate class are boys.

Number of boys = 40/100 x 40 = 16

Number of Girls = 40 – 16 = 24

Let the number of girls who joined be X and the % of girls is 80%

24 + X = 80/100 x (40 + X)

24 + X = 32 + 4X/5

X/5 = 8

X = 40

Solution to Question 102

Let the total money be $X.
James got 20% = 0.2X
Alice got = $2800
William got = 60/100 x 2800 = 1680
X = 0.2X + 2800 + 1680
0.8X = 4480
X = 4480/0.8 = 5600
The sum of money taht was shared among 3 childre = $5600

Solution to Question 103

Let the money spent by Andy be X
Penny spent 20% less. That means she spent 0.8X
X + 0.8X = 1980
1.8X = 1980
X = 1980/ 1.8 = $1100
Money spent by Andy = $1100

4. Ratio and Proportion

Solution to Question 104

The scale used in making the model of the building is 2:450.
That means a 450 in real life will be 2 in the model.
If the building is 81 meters high, then the height of the model will be:
81/450 x 2 meters.
To get the height of the building in centimeters, we will multiply it by 100 to get:
81/450 x 2 x 100
= 36 centimeters

Solution to Question 105

There are 32 green markers.
The number of red markers = 32 x 5 = 160
Let the number of green markers to be added be X.
(32 + X) / 160 = 3/4
4 x (32 +X) = 3 x 160
128 + 4X = 480
4X = 480 – 128 = 352
X = 352/4 = ~~98~~ 88
The number of green markers that should be added = 98 green markers.

Solution to Question 106

a) It is given that for every 15 cakes ordered, 5 were cheese cakes and 10 were mango cakes.
Therefore the ratio of cheese cakes to mango cakes = 1 : 2
Let the number of cakes be X
Then the number of cheese cakes sold will be X/3 and the number of mango cakes sold will be 2X/3
30 x X/3 + 36 x 2X/3 = 3060
10X + 24X = 3060
34X = 3060
X = 3060/34 = 90 cakes
Total number of Cheese cakes sold = 1/3 x 90 = 30
b) Number of Mango cakes = 2/3 x 90 = 60
Money obtained from selling all Mango cakes = 60 x 36 = $2160

Solution to Question 107

Let the second number be X
Then the first number is = 3X/8
The difference between the two numbers is 45
X – 3X/8 = 45
8X – 3X = 45 x 8
5X = 360
X = 360/5
X = 72
The value of the second number = 72

260

Solution to Question 108

Let the total number of people at the party be X
The number of men = 3X/10
And the number of women = 7X/10
There are 36 more women than men.
3X/10 + 36 = 7X/10
7X/10 − 3X/10 = 36
4X/10 = 36
X = 36 x 10/4 = 90
No of people at the party = 90

Solution to Question 109

18 years ago, if Carl's age was 7X, then David's age would be 2X (That is why the ratio is 7X : 2X which is equal to 7 : 2)
Today (18 years later), their ages are 7X + 18 and 2X + 18
The ratio now is 2:1. So
(7X + 18)/ (2X + 18) = 2/1
7X + 18 = 2 x (2X + 18)
7X + 18 = 4X + 36
7X − 4X = 36 − 18
3X = 18
X = 6
a) Carl's age now is 7X + 18 = 7 x 6 + 18 = 42 + 18 = 60 years
David's age now is 2X + 18 = 2 x 6 + 18 = 12 + 18 = 30 years
b) David's age18 years from will be 30 + 18 = 48 years.

261

Solution to Question 110

Let the number of shells Richard collected be X.
Shells collected by William = X + 40
Shells collected by Nancy = 1/4 x (X+40)
X + 70 = 1/4 x (X + 40) + X + 40
4X + 280 = X + 40 + 4X + 160
X = 80
Number of shells William collected = 120

Solution to Question 111

Let the number of Chinese at the party be X.
Number of Indians = 25X/100 = X/4
24 Chinese and 13 Indians joined the party and the ratio was 3:1

$$\frac{X + 24}{X/4 + 13} = \frac{3}{1}$$

Cross multiplying we get
X + 24 = 3X/4 + 39
4x + 24 x 4 = 3X + 39 x 4
X = 156 – 96 = 60
The number of Chinese at the party initially = 60

Solution to Question 112

Let the savings of James be 9X and the savings of Margaret be 2X
After spending $80 and $30, the new ratio is 10:1

$$\frac{9X - 80}{2X - 30} = \frac{10}{1}$$

9X – 80 = 10x (2X – 30)
9X – 80 = 20X – 300
300 – 80 = 20X – 9X

262

220 = 11X
X = 220/11 = 20
The amount of money James had at first = 9x20 = $180
The amount of money with Margaret at first = 2 X 20 = $40
After spending $30, she is left with $40 – $30 = $10

Solution to Question 113

Let the total number of people at a gathering be X.
Number of adults = 4X/9
Number of children = 5X/9
When 6 children left, ratio became 8 : 7

$$\frac{4X/9}{5X/9 - 6} = \frac{8}{7}$$

Cross multiplying we get
28X/9 = 40X/9 – 48
40X/9 – 28X/9 = 48
12X/9 = 48
X = 48 x 9/12 = 36
Number of Children in the gathering at first = 5/9 x 36 = 20

Solution to Question 114

Initially number of beads = 180
Number of Blue beads = 5/9 x 180 = 100
Number of Red beads = 4/9 x 180 = 80
Some beads are added to the container and the total number of beads = 300 and the new ratio is 3:2
New number of blue beads = 3/5 x 300 = 180
New number of red beads = 2/5 x 300 = 120
a) Number of blue beads added = 180 – 100 = 80
Number of red beads added = 120 – 80 = 40
b) 40 more blue beads are added than red beads

Solution to Question 115

Let the initial number of tarts be 2X, 3X and 5X
After 12 new tarts of each kind the ratio became 7:9:13
(2X+12): (3X+12) = 7:9
9 x (2X + 12) = 7 x (3X + 12)
18X + 108 = 21X + 84
3X = 24
X = 3
The number of apple tarts to begin with are 3X = 3 x 8 = 24
After making 12 more tarts, the number of apple tarts = 24 + 12 = 36

Solution to Question 116

Let the total number of apples in baskets A & B together be X
There are 98 apples in C.
The ratio of the total number of apples in A and B to the number of apples in C
is 4:2

$$\frac{X}{98} = \frac{4}{2}$$

X = 98 x 4/2
X = 196
Now the number of apples in A and B are in the ratio 3:4
That would mean the number of apples in A are 3/7 of the total apples in A and B
The number of apples in Box A = 3/7 x 196
= 3 x 28
= 84
The number of apples in Basket A = 84

Solution to Question 117

Let the total no of books be X

Number of story books = 6X/8

Number of art books = 2X/8

By the end of the week, 25% of each type of books were sold and 1620

books of both types were unsold.

Remaining books were 75% of the initial number of books

75X/100 = 1620

X = 1620 * 100/75 = 2160

Number of story books = 6/8 x 2160 = 1620

Number of art books = 2/8 x 2160 = 540

Solution to Question 118

The marbles thta Rachael and John have are in the ratio 2:3

Rachel has = 2/5 x 80 = 32

John has = 3/5 x 80 = 48

Let the number of lost marbles be X.

Number of marbles remain with John = 48 − X

It is given that the marbles John had remaining, made up 1/3 of the total number

of marbles they both had left.

(32+48 − X)/3 = 48 − X

80 − X = 144 − 3X

3X − X = 144 − 80

2X = 64

X = 32

Therefore number of marbles John had in the end = 48 − 32 = 16 marbles

Solution to Question 119

Let the total number of Vehicles in the car park be X
The number of cars and vans were in the ratio = 1:2
Cars = X/3
Vans = 2X/3
68 cars entered and 18 vans left, then the ratio became 7:3

$$\frac{X/3 + 68}{2X/3 - 18} = \frac{7}{3}$$

Cross multiplying we get
X + 68 x 3 = 14X/3 − 18 x 7
3X + 204 x 3 = 14X − 126 x 3
14X − 3X = 612 + 378
11X = 990
X = 990/11 = 90
Therefore number of cars = 90/3 = 30
Number of vans = 2 x 90/3 = 60

5. Area, Perimeter and Volume

Solution to Question 120

1st circle's area = π x 72 = 49 π
2nd circle's area = π x 142 = 196 π
Therefore ratio = 49 : 196 = 1 : 4

Solution to Question 121

Volume of the rectangular tank = 65 x 40 x 35 = 91000 cubic cm
Volume when it is 1/2 full = 0.5 x 91000 = 45500 cubic cm
1 cubic cm = 0.001 Liters
45500 cubic cm = 45.5 Liters
The water was then used to fill some bottles completely.

266

The capacity of each bottle was 1.5 liters.No of bottles that can be filled completely = 45.5/1.5 = 30
The amount of water left in the tank when all the bottles were filled completely
= 45.5 − 30 x 1.5
=45.5 − 45
= 0.5 Liters

Solution to Question 122

Volume of the rectangular box = 5 x 4 x 2 = 40 cubic cm
Volume of the cubic box = 10 x 12 x 10 = 1200 cubic cm
The greatest number of the blocks that can fit in a cubic box = 1200/40 = 30

Solution to Question 123

The tank contains $720m^3$ of water when filled to the 4 m mark.
Let the length and breadth of tank be L, B
L x B x 4 = 720
L x B = 180 m^2
The capacity of the tank = L x B x H
H = 10 m.
The capacity of the tank = 180 x 10 = 1800 m^3

Solution to Question 124

The volume of water in both the tanks will be the same.
Volume of tank with full of water = 10 x 4 x 4 = 160 cubic meters
Let the height of the second tank be 'H'
Volume of 2nd tank = 8 x 4 x H = 160 cubic meters
32H = 160
H = 160/32 = 5 meters
Height of water in the second tank is 5 meters.

Solution to Question 125

The figure has 4 quarter circles which makes a complete circle.

Perimeter of a circle = d x π

$$= 14 \times 22/7 = 44$$

Perimeter of the following figure = 16 + 16 + 44

$$= 76 \text{cm}$$

Solution to Question 126

For any regular polygon with n sides, the sum of all the internal angles is =
(n − 2) x 180°

n = 6 for a hexagon.

The sum of all internal angles = (6 − 4) x 180 = 4 x 180 = 720°

Solution to Question 127

Area of the square = 28 x 28 = 784 sq cm

2 quarter circles make 1 semi circle

Area of the shaded part = area of the quarter circle ACB + Area of the quarter circle ACD − Area of the Square

Area of a semi circle = $\prod r^2/4$ = 22/7 x 28 x 28 x ¼ = 616 sq cm

Area of the shaded region = 2 x 616 − 784 sq cm

= 448 sq cm

Solution to Question 128

Perimeter of the triangle = 6 + 7 + 8 = 21cm

Perimeter of a rectangle is twice that of a triangle .

Perimeter of the rectangle = 21 x 2 = 42

now perimeter of a rectangle = 2 (length + breadth)

2 (L + B) = 42

Length + Breath = 21

Since each value is a whole number, the possible values are

Length	Breadth
1	20
2	19
3	18
4	17
5	16
6	15
7	14
8	13
9	12
10	11

If we go beyond 10, the values of length and breadth will be reversed. So the number of different possible values is 10

Solution to Question 129

Capacity of the rectangular tank = 14 x 20 x 42 = 11760 cubic cm
1 cubic cm = 0.001 Liters
11760 cubic cm = 11.76 Liters
The tank is 3/4 full = ¾ x 11.76 = So it has 8.82 Liters of water
The water was then released through a hole at a rate of 1 liter per minute. After 5 minutes, the loss of water = 1 x 5 = 5 Liters
Remaining water in the tank = 8.82 − 5 = 3.82 Liters
Remaining capacity of the tank = 11.76 − 3.82 = 7.94 Liters
Fraction of the tank that was not filled with water = 7.94/11.76 = 397/588

Solution to Question 130

Each rectangle has a perimeter of 40 cm.
Let the length and breadth of the rectangle be L & B
2 x (L + B) = 40
L + B = 40/2 = 20
The square photo frame is made in a way that the side of the square is equal to the length of one rectangle and the breadth of the other rectangle.
Since all the rectangular pieces are identical, the side of the square = (L+B).
Area of the square = (L+B)2
= 202
= 400 cm2

Solution to Question 131

Count the no of cubes along the length, breadth and height which gives you the

Length = 10 cm

Breadth = 8 cm

Height = 3 cm

Volume of this rectangular prism = 10 x 8 x 3 = 240 cm^3

Solution to Question 132

Volume of the box = 9 x 6 x 4 = 240 cubic cm

Volume of the cube = 2 x 2 x 2 = 8 cubic cm

Number of cubes of side 2 centimeters each you can put into this box = 240/8 = 30

Solution to Question 133

Volume of the tank = 60 x 40 x 25 = 60000 cubic cm

1 cubic cm = 0.001 Liters

60000 cubic cm = 60Liters

3/4 of the glass tank is filled with water = 60 x ¾ = 45 Liters

Ben transfers some of the water from the glass tank to a steel tank until it is filled to its brim .

10 liters of water is left in the glass tank.

Therefore water transferred to the steel tank = 45 – 15 = 30 Liters

The steel tank is 25cm long and 20cm wide.

Let the height of steel tank be H.

25 x 20 x H = 30000

H = 30000/500

H = 60 cm

The height of the steel tank = 60 cm

 **Solution to Question 134**

a) Perimeter of the figure = perimeter of the 4 quarter circles with radius = 14 cms
(the side of square = 28 cms and radius = half the side)
Perimeter = 4 x (2 x 22/7 x 14 / 4) = 4 x 22 = 88 cms
b) Area of the shaded region = area of the square − area of the 4 sectors
= 28 x 28 − 4 x (22/7 x 14 x 14 /4)
= 28 x 28 − 22 x 28
= 168 sq cms

 Solution to Question 135

Length of the rectangle is twice its breadth.
Length = 14cm, therefore breadth = 14 /2 = 7 cm
Circumference of circle = 2rΠ ; area = Π r^2
2 semi circle's makes 1 complete circle
large circle's radius = 14/2 = 7 cm
Small circle's radius = 7/2 = 3.5 cm
Perimeter of the large circle = 2 x 22/7 x 7 = 44 cm ;
area = 22/ 7 x 7 x 7 = 154 sq cm
Perimeter of the small circle = 2 x 22/7 x 3.5 = 22 cm; area = 22/7 x 3.5 x 3.5 = 38.5 sq cm
Perimeter of the fig = 44 + 22 = 66 cm
Area = 154 + 38.5 = 192.5 sq cm

 Solution to Question 136

Let the breadth of the rectangle be X.
Length = 140X/100
Perimeter of a rectangle = 2x(L + B)
2 x (140X/100 + X) = 144
140X + 100X = 144 x 100/2
240X = 7200
X = 7200/240 = 30
Breadth of the rectangle = 30cm
Length of the rectangle = 140/100 x 30 = 42cm

Let the side of the square be X.

Area of the square = X x X = X^2

Radius of circle = X/2

Area of semi circle CD = π x (X/2)^2/2 = $\pi X^2/8$

If we subtract the area of semi circle's CD & AB from the square, then we will get the middle part area. (Refer figure)

$X^2 - 2x\pi X^2/8$

$X^2 - \pi X^2/4$

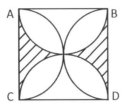

If we double the area, we get the total shaded area in the figure

= 2 x ($X^2 - \pi X^2/4$) = $2X^2 - \pi X^2/2$

Subtract this area from the area of the square, to get the area of the 4 leaves.

$X^2 - (2X^2 - \pi X^2/2)$

= $\pi X^2/2 - X^2$

Volume of the tank = 90 x 50 x 30 = 135000 cubic cm

30% is filled with oil that means 70% of the tank is empty

Volume of the empty part = 0.7 x 135000 cubic cm.

The time taken to fill the tank at 900 cm^3 per minute = 0.7 x 135000 / 900

= 105 min

The total time taken by the tap to fill the tank to its brim = 105 min

= 1 hr 45 min

272

Solution to Question 139

a) Let the tank's capacity be X liters

Given that 2 taps fill 6 liters water each per minute

Total water filled per minute is 2 x 6 =12 liters

Water drained out at a rate of 8 liters every 4 min , In 1min water drained = 8/4=2 Liters

Net water filled per min 12 − 2 = 10 liters per min.

In ten minutes water filled in the tank = 10 x 10 = 100 liters

Given that only 40% of the tank is full after 10 minutes,

ie. X x 40/100 = 100 liters

X = 100 x 100/40 = 250 liters

Capacity of the tank was 250 Liters.

b) 40% of the tank was filled in 10 minutes.

The remaining 60% will get filled in 10/40 x 60 = 15 minutes

Solution to Question 140

Length of the garden = 24m.

The perimeter of the garden is = 78m.

If B is the breadth of the garden, then 2 x (24 + B) = 78

24 + B = 39

B = 39 − 24 = 15m

Area of the piece of land = 24 x 15 = 360 sq meters

Now 60% of the land was used to make the school. So 40% of the land remained after that.

One fourth of this land = ¼ x 40% = 10% was used to make the garden.

Since area of the garden is 360 square meters and this is 10% of the original piece of land, the area of the original land = 100%/10% x 360 = 10 x 360 = 3600 square meters.

Area of the school = 60% x 3600 = 2160 square meters.

Difference between the area of the school and the area of the garden = 2160 − 360 = 1800 square meters.

The area used to build the school was 1800 square meters bigger than the area of the garden.

Solution to Question 141

Area of rectangular base = 360 sq cm

Height of the rectangular tank is given as 15cm

Volume of the tank = 360 x 15 = 5400 cubic cm

1200 cubic cm is poured away

Remaining water left = 5400 – 1200 = 4200 cubic cm

This remaining water is poured into 2cm cubes.

Volume of the cube = 2 x 2 x 2 = 8 cubic cm (all the sides of a cube are equal)

Number of cubes required to pour the remaining water = 4200 / 8 = 700

Solution to Question 142

Let the capacity of the reactangular can be X liters.

Grace poured out 1.62 liters of water from a rectangular can, the volume of water was reduced by 1/5

This means X/5 = 1.62

X = 1.62 x 5 = 8.1 liters

Capacity of the can = 8.1 liters

Let the breadth of the rectangular base of the can be K cms

Length exceeds its breadth by 6cm, so Length = K + 6

Perimeter = 2 x (L + B) = 0.6m = 60 cms

2 x (L+B) = 60

X + X + 6 = 30

2X = 24

X = 12 cms = 0.12m

Breadth = 12cm, Length of the rectangle = 12 + 6 = 0.18 meters

Let height be H meters.

Volume of the can = 12 x 18 x H = 8,1 liters

0.12 x 0.18 x H = 0.0081

H = 0.375 meters

Depth of water in the can if it was filled to its capacity originally = 0.375 meters

Solution to Question 143

Area of bigger circle = $\prod r^2$ = 22/7 x 7 x 7 = 154 sq. cms
Area of smaller circle =22/7 x 3.5 x 3.5 = 38.5 sq. cms
Area of 2 small circles = 2 x 38.5 = 77
Area of shaded region = 154 – 77 = 77 sq. cms

Solution to Question 144

a) Length of small cuboids = 4cm
Length of rectangular container = 24 cm
Number of cuboids fit with this length = 24 / 4 = 6
Width of small cuboids = 22cm
Width of rectangular container = 3cm
number of cuboids fit with its width = 22/3 = 7
Height of small cuboids = 2 cm
Height of rectangular container = 19 cm
number of cuboids fit in this height = 19/2 = 9
 Therefore total number of cuboids that can fit in = 6 x 8 x 7 = 336
b) Number the cubes will not fit the whole container exactly.
Volume of container remains empty
= 24 x 22 x 19 – 336 x 67(4 x 3 x 2)
= 10032 – 8064 = 1968 cubic cm

Solution to Question 145

Volumes of Cube A adn Cube B are in the ratio 2:1   432
So if the volume of cube A is 216 cm^3, the volume of cube A will be 5̶1̶2̶/2 = 216 cm^3
Let each side of the cube be A cms.
Then A x A x A = 216
A = 6 cms
The length of a side of cube B = 6 cms.

Solution to Question 146

a) Let the volume of tank be X liters

At the end of 6 min, the tank is 2/3 full

That means 1/3 of tank got emptied = X/3

So the rate of flow is X/3/6 = X/18 liters per minute

After a further 8 minutes 2.4 liters apple juicewas left. That means X – 2.4 liters had flown out in 14 minutes.

X/18 x 14 = (X – 2.4)

14X = 18X – 18 x 2.4

18 x 2.4 = 18X – 14X = 4X

X = 18 x 2.4/4 = 10.8 liters

Volume of tank is 10.8 Liters

b) The amount of juice that flowed out in 1 minute = X/18 liters

= 10.8/18 = 0.6 liters per minute

Solution to Question 147

Using a pin, a string and a pencil, the largest possible shape Thomas can draw is a circle.

The length of the string will be the radius of the circle = is 70mm long = 7cm

Area of the circle of radius 7 cms = 22/7 x 7 x 7 = 154 sq cms

Solution to Question 148

a) Sum of the interior angles of a regular polygon = 180 x (n – 2) where n is the number sides.

b) Sum of the angles of a Decagon = 180 x (10-1) = 180 x 9 = 1620

Solution to Question 149

Volume of the tank = 15 x 20 x 30 = 900 cubic cm

It was already 1/5 filled with water = 1/5 x 9000 = 1800 cubic cm

Volume of remaining tank to be filled = 9000 − 1800 = 7200

Total water that flowed from both the taps = 60 + 90 = 150 ml/min

Time taken to fill the remaining tank = 7200/150 = 48 min

If both taps fill the container at a rate of 100 ml/min = 100 + 100 = 200 ml/min

Time taken = 7200/200 = 36 min

It will take 12 min faster to fill the tank if both taps fill the tank at a rate of 100m/min

6. Speed, Distance and Time

Solution to Question 150

a) First calculate the distance between Singapore and Kuala Lumpur with the help of car data.

Distance = speed x time

Distance = 90 x 5 = 450km

b) Speed of the bus = 450 / 9 = 50 km/hr

c) 3/5 of the journey is completed and the remaining journey = 2/ 5 x 450 = 180km

Actual time taken for 180 kilometers = 180 / 50 = 3.6 hr = 3 hr 36 min

But the bus needs to reach earlier in 1 hr 36 minutes

Total time needed to reach = 3 hr 36 min − 1 hr 36 min = 2 hr

Speed = 180 /2 = 90 km/hr

The bus should travel at the speed of 90km/hours after 3/5 of the journey.

 **Solution to Question 151**

a) Distance = speed x time
Distance Francis jogged = 8 x 1 = 8 km
Distance both Francis and Christopher jogged = 8 kilometres
b) Christopher started jogging at 2:30 along the same path and met Francis at 3:15
Time taken by Christopher = 3:15 − 2:30 = 45 min
Christopher's jogging speed = 8 / (3/4) = 32/3 kilometres/ hr
c) Difference in their speeds = 32/3 − 8 = 8/3 kilometres/ hour
That is the ditance by which they will be apart in 1 hour
To be 6 kms ahead, time required by Christopher = 6 / (8/3) = 9/4 hours = 2 hours 15 minutes
Christopher will be 6km ahead of Francis in 2 hours 15 minutes

 Solution to Question 152

Time taken by Shawn = 5:30 − 1:30 = 8 hr
Distance = 90 x 8 = 720 kilometers
Mary started driving at the same time and reached her destination 6 hours later.
Total time taken by Mary = 8 + 6 = 14hr
Mary's speed = 720/14 = 51.42km/hr

 Solution to Question 153

In one day Johnson can make 360 kites.
Debby can make 2/3 as many kites as Johnson = 2/3 x 360 = 240
In 1 day both can make = 360 + 240 = 600 kites
In one day they can make 600 kites.
So 3600 kites can be made in:
3600/600 = 6 days

 Solution to Question 154

a) A carpenter takes 6 minutes to cut a piece of wood into 2 pieces.
To cut into 5 pieces he will need 4 cuts
Time required = 4 x 6 = 24 min
b) For 30 pieces, he will need to cut 29 times
Time taken for 30 pieces = 29 x 6 = 174 min = 2 hr 54 min

278

 **Solution to Question 155**

a) Total time taken by motorist A = 6 hrs
Distance traveled = 80 x 6 = 480 km
Motorist B left the town X 2 hrs after the motorist A
= 11:30 am
Time taken by motorist B = 480/100 = 4.8 hr
b) Motorist B reached Town Y at 3:30 pm
Motorist B left Town X at 11:30 am
Therefore time taken = 4hr
Speed = distance / time = 480/4 = 120 km/hr
Motorist B must travel with a speed of 120 km/hr in order to reach Town Y at the same time as motorist A.

 Solution to Question 156

a) Let the distance of the total journey be X kilometers
Jason completed the remaining 4/9 of his 24km journey in 12 minutes.
Hence 4/9 x = 24
x = 24 x 4/9 = 54 km
First part of the journey = 54 – 24 = 30 km
Time taken = 30/80 hours = 22.5 minutes
Total time taken = 22.5 min + 12 minutes = 34.5min
b) For second part of the journey
Distance = 24 kilometers
Time = 12 min = 0.2 hours
Speed for the second part of the journey = 24/0.2 = 120 kilometers/ hr

 Solution to Question 157

Sam drove at an average speed of 84 km/h for 45 min.
Distance = 84 x 0.75 = 63 kilometers
He then reduced his average speed by 6 km/h and drove a further 30 min
Speed = 84 – 6 = 78 km/hr
Distance = 78 x 0.5 = 39 kilometers
Total distance covered by Sam = 63 + 39 = 102 km

279

Solution to Question 158

Let us assume that the number of hours required for Car A to overtake Car B be X.

Since car B left 3 hours earlier, in the same time it would have travelled for (X +3) hours.

When car A overtakes car B, the distance travelled by both is the same.

Distance = speed x time

60 x (X + 3) = 80 x X

60X + 180 = 80X

180 = 80X − 60X

180 = 20X

X = 180/20 = 9 hours

So the time when Car A overtakes Car B = 11 AM + 9 hours = 8 .00 PM

Solution to Question 159

Time difference between the starting time of the cyclist and the motorist = 4 hours.

The motorist took 4 hr to catch the cyclist.

Total time that the cyclist has driven = 4 + 4 = 8 hr

Distance = 60 x 8 = 480 kilometers

Average speed of the motorist = 480 / 4 = 120 km/hr

Solution to Question 160

The distance between Bangkok and Kuala Lumpur was 720 km.

Andy took 6 hour to travel from Bangkok to Kuala Lumpur.

Speed of Andy = 720 / 6 = 120 km/hr

Gavin left Bangkok 30 min earlier than Andy and traveled at 20 km/h slower than Andy.

Gavin's speed = 120 − 20 = 100 km/hr

Distance travelled by Gavin in 6 hrs = 100 x 6 = 600 kms

Gavin was 720 − 600 = 120 km far away from Kuala Lumpur when Andy reached.

Solution to Question 161

a) Time taken to travel from Marina Bay and Anderson Point = 1:45 − 9:15 = 4.5 hr
Distance between Marina Bay and Anderson Point = 4.5 × 80 = 360 km
A lorry driver left Marina Bay 30 minutes earlier but reached Anderson Point 1 hour later than the motorist
Time taken by lorry driver = 4.5 + 1.5 = 6 hr
Speed = 360/6 = 60 km/hr
If he wants to reach the Anderson Point at the same time as the motorist, he will have to reach one hour earlier. So his travel time = 6 − 1 = 5 hr
Speed = 360/5 = 72 km/hr
b) The lorry driver should increase his speed by 72 − 60 = 12 km/hr to reach Anderson Point at the same time as the motorist.

Solution to Question 162

Average speed = total distance / total time
Total distance between Bay Front and the New Town = 468
For 2 hr Charles maintained a speed of 72km/hr
Distance traveled = 72 × 2 = 144 kms
Remaining journey = 468 − 144 = 324 kms
Charles' new speed = 72 − 18 = 54 kms/ hour
Time taken to complete the remaining journey = 324/54 = 6 hrs
Total time for the journey = 2 + 6 = 8 hours
Average speed   = total distance/ total time
                  = 468/8 = 58.5 kms/ hr
Average speed of Charles was 58.5 km/hr.

Solution to Question 163

Time taken by Williams = 36/48 = 0.75 hr = 45 min
Last Sunday, he left his home 9 minutes earlier and still arrived at the school at the usual time.
Time taken = 45 min + 9 = 54 min = 0.9 hr
Distance = 36 kms
New speed = 36/0.9 = 40 km/hr
He decreased the speed of his motorcycle by = 48 − 40 = 8km/hr

Solution to Question 164

Let the point where Margaret and Alicia meet be X kms from Victoria Central.
That point will be 900 – X kms from Pebble Bay
Time taken by Margaret = X/80
Time taken by Alicia = (900 – X)/70
Since they left at the same time, the time taken by both to reach their meeting point is the same
X/80 = (900 – X)/ 70
70X = 80 x 900 – 80X
150X = 80 x 900
X = 900 x 80 / 150 = 480 Kms
Margaret covered a distance of 480kms when she met Alicia.

Solution to Question 165

The given situation is in inverse proportion
Let the number of workers be X.

$$\frac{40}{X} = \frac{20}{8}$$

Cross multiplying we get
40 x 8 = 20X
X = 320/20 = 16
16 workers are required to paint a building in 20 days.

Solution to Question 166

Time taken to travel from port of Eden to Port of Stephens = 900/150 = 6 hours
Time taken to Newcastle port = 600/120 = 5 hours
Total time for the whole journey = 6 + 5 = 11 hours

Solution to Question 167

Tiger airways plane's speed = 324 km/hr
Singapore airlines plane's speed = 360 km/hr
Let the time taken by the tiger airway plane be X hours.
Hence the time taken by Singapore Airlines will be X – 1 hours.
When the two flights meet, the distance traveled by both will be equal.
324 x X = 360 x (X – 1)
324X = 360X – 360
360X – 324X = 360
36X = 360
X = 360/36 = 10 hrs
The distance traveled by Tiger airways plane before Singapore airlines plane
caught up =
10 x 324 = 3240 kms

Solution to Question 168

Jason's speed = 60 km/hr
Distance travelled by Jason in 2 hr = 60 x 2 = 120 km/hr
After traveling 2 hrs Jason and Gavin are 260 kilometres apart
Therefore the distance traveled by Gavin = 260 – 120 = 140 kilometres
Average speed of Gavin = 140/2 = 70 km/hr
Average driving speed of Gavin as 70 km/hr.

Solution to Question 169

Let the round trip to Sydney be X weeks long
Hence the trip to Beijing is 2X weeks long
The trip to London is 4X weeks long
The three round trips take 8 weeks to complete
X + 2X + 4X = 8 x 7 days (1 week = 7 days)
7X = 8 x 7
X = 8 days
No of days for a one-way trip to Beijing = 8 days

Solution to Question 170

The motorist took 4 hour to catch up with the cyclist
Time for which the cyclist has travelled till then = 8 hours
Total Distance travelled by the cyclist in 8 hr = 60 x 8 = 480 kilometers = distance travelled by the motorist
Speed of motorist = 480/4 = 120 km/hr
Average speed of the motorist was 120km/hr.

Solution to Question 171

Total distance from Town A to Town B = 560 km
Caroline left Town A at 7.15 am and drove towards Town B along the same expressway as Josh.
She passed him at 10.15 am
Time taken by Josh till they meet = 10:15 – 6:45 = 3.5 hr
Distance travelled by Josh in 3.5 hr = 60 x 3.5 = 210 kilometers
Time taken by Caroline to cover that distance = 10:15 – 7:15 = 3 hr
Speed of Caroline = 210/3 = 70 km/hr
Caroline continued driving at the same speed until she reached Town B
Total time taken by Caroline for travelling 560 km = 560/70 = 8 hrs
Distance traveled by Josh in 8 hr = 60 x 8 = 480 kilometers
Distance by which Josh was away from town B when Caroline reached Town B = 560 – 480 = 80 kms

Solution to Question 172

Bees	time	jars
2000	1yr	7
5000	?	70

As the number of bees increases the number of jars also increases. So it is in direct proportion.
2000 bees make 7 jars in one year, so 5000 bees will make
= 5000 x 7 /2000 = 17.5 jars in 1 year
Number of years required for 70 jars = 70/17.5 = 4 years.
It will take 4 years for 5000 bees to make 70 jars of honey.

Solution to Question 173

On the last day the earthworm will travel 3.5 meters.

So before that it will need to travel 6 – 3.5 = 2.5 m

In a day and a night the earthworm travels a net of 1.5 meters. So in 2 days and 2 nights it will travel 3 meters

On the third day, the earthworm will reach 3 + 3.5 meters = 6.5 m and will be out of the well.

Solution to Question 174

Average Speed = Total Distance / Total Time

Let the distance between London to Amsterdam be X

Distance = speed x time

Time taken for forward journey = X/300

Time taken for backward journey = X/600

Total time taken for round trip = X/300 + X/600 = (2X + X)/600 = 3X/600

Total distance = 2X

Average Speed for the round trip = 2X/3X/600 = 2X x 600/3X = 400 km/hr

Solution to Question 175

Katie and Jim started walking to school at 8:00 AM at 4 kilometers per hour.

10 minutes later, Katie realized that she left her homework notebook at home.

Distance traveled by both in 10 min = 2/3 km

If Jim now runs towards home and comes back and catches his sister in T hours, the distance travelled by Jim =

8 x T (His speed is 8 kms/ hour)

The extra distance travelled by Katie in that much time = 4 x T kms

8T = 2/3 + 2/3 + 4T

8T – 4T = 4/3 hours

4T = 4/3 hours

T = 1/3 hours = 20 mins ⟶ So Jim will take 20 minutes to catch up with Katie

He started running towards home at 8:10 AM, so the time at which he will catch Katie = 8:30 AM

Solution to Question 176

Let the speed of cat be X.
Speed of the dog = 2X
Distance = speed x time
Time taken by the dog to travel 1000m = 1000/2X
Distance traveled by the cat in 1000/2X time = X x 1000/2X = 500m
The distance of the cat from home, when the dog reached home = 800 – 500 = 300meters.

Solution to Question 177

a) Let the age of Ben be X.
Ms. Rachel is five times as old as her son, so her age = 5X
In 8 years' time, their total ages will be 58 years.
X + 8 + 5X + 8 = 58
6X = 58 – 16
X = 42/6 = 7 years
Kevin is twice as old as Ben now = 2 x 7 = 14 years
Ben's present age = 7 years
b) Kevin's present age = 14 years
Ms. Rachael's present age = 35 years

7. Time, Age and Money

Solution to Question 178

A bus and a car leave the same place and travelled in opposite directions.
The bus is travelling at 50 km per hour and the car is travelling at 55 kilometers per hour.
Total distance travelled by them in 1 hour = 50 + 55 = 105 km
Remaining journey = 200 – 105 = 95 kms
1 hr ——— 105 km
? Hr ——— 95 km
= 95/105 = 0.9 hour i.e. 54 min
Number of hours in which they will be 200 kilometers apart = 1 hour 54 minutes

Solution to Question 179

Let the number of years be X after which Susan will be thrice as old as Nicolas.
3 x (4 + X) = 24 + X
12 + 3X = 24 + X
3X − X = 24 − 12
2X = 12
X = 6
After 6 years, Susan will be thrice as old as Nicolas.

 ## Solution to Question 180

When Gavin is 15 years old, his sister is 8 years old and his father is 45 years old.
Let us assume that after X years, Gavin is half his Father's age
After X years,
Gavin's age = 15 + X
Father's age = 45 + X
Gavin is half his father's age
2 (15 + X) = 45 + X
30 + 2X = 45 + X
2X − X = 45 − 30
X = 15
After 15 years, Gavin will be half his Father's age.
Age of Gavin's sister when Gavin is half his father's age = 8 + 15 = 23 years

Solution to Question 181

Let the age of the youngest sibling be X years.
Then the age of the other siblings will be:
X+ 3, X + 6, X + 9 and X + 12 years.
The sum of their ages is 70 years.
So:
X + X+ 3 + X + 6 + X + 9 + X + 12 = 70
5X + 30 = 70
5X = 70 − 30 = 40
X = 40/5 = 8 years
The youngest sibling is 8 years old

Solution to Question 182

If Jim is 35 years old this year, then 3 years ago, his age was:
35 − 3 = 32 years.
When he was 32, he was 4 times as old as Claire was then.
Age of Claire when Jim was 32 = 32/4 = 8 years.
Claire's age this year = 8 + 3 = 11 years.

Solution to Question 183

Let James' current age be X years.
When he was born, his fathers age was 3X years.
So his fathers current age is 3X + X = 4X years.
His grand father, Mr. Smith's age is 8X years.
Mr. Smith was 32 when his son (James' father) was born.
So Mr. Smith is 32 years older than his son.
This means:
8X = 4X + 32
8X − 4X = 32
4X = 32
X = 32/4 = 8 years old.
James' current age is 8 years

Solution to Question 184

Fenny's father's age in 1995 = 33 years
In 2004 fenny's fathers age = 33 + 9 = 42 years
Fenny's age = 2/7 x 42 = 12 years
Let X be the the number of years after which Fenny is 5/11 of her father's age.
Father's age = 42 + X
Fenny's age = 12 + X
5/11 x (42 + X) = 12 + X
210 + 5X = 132 + 11X
11X − 5X = 210 − 132
6X = 78
X = 78/6 = 13 years
Therefore year = 2004 + 13 = 2017
Year in which Fenny's age will be 5/11 of her father's age will be 2017.

Solution to Question 185

Let Mary's age be X,
Mary's father's age = 4X
a) Mary's brother's age = 2X
Mary's Father was 36 years old
4X = 36
X = 9
Mary's age = 9 ,Mary's brother's age = 18 years
Age of Mary's brother in 2000 i.e. after 2 years = 18 + 2 = 20 years
In 2003 Mary's age = 9 +5 = 14 years
Her brother's age = 18 +5 = 23 years
Her father's age = 36 + 5 = 41 years
b) Total sum of the age of the family = 14 + 23 + 41 = 78 years

Solution to Question 186

Carl paid 20% of the cost of Television
Cost of Television = $1500
Down payment = 20/100 x 1500 = $300
Remaining amount to be paid in 12 monthly installments = 1500 – 300 = 1200
Amount to be paid each month = 1200/12 = $100

Solution to Question 187

Let the selling price of the computer be $X
A computer was sold at $1280 after a 20% discount.
This means 80X/100 = 1280
X = 1280 x 100/80 = $1600
Discount = 1600 – 1280 = $320
 When Susan bought the computer, she was given a further 15% discount on the selling price.
Discount = 1600 x 15/100 = $240
If Susan had bought the computer without discount, Susan would need to pay more = 320 + 240 = $560 more
Susan would have to pay $560 more if she had bought the computer without discount.

289

 ## Solution to Question 188

The price of a furniture in January = $1500
The price was increased by 12% in February. 1500 x 12/100 = $180
Price in February = 1500 + 180 = $1680
In March ,it was decreased by 30% = 1680 x 30/100 = $504
Price in March = 1680 – 504 = $1176
The difference in the price of furniture in January and in March
= 1500 – 1176 = $324

 ## Solution to Question 189

Original price of the camera = $900
20% discount = 900 x 20/100 = $180
Therefore Jacob bought camera for 900 – 180 = $720
Camera + laptop = $3520
The amount paid for Laptop = 3520 – 720 = $2800

 ## Solution to Question 190

a) Total number of kittens and turtles = 60
Let the number of kittens be X
number of turtles = 60 – X
Each kitten cost $5. Each turtle cost $3. If the total cost of the kittens was $100
more than the turtles
3 x (60 – X) + 100 = 5X
180 – 3X +100 = 5X
5X + 3X = 280
8X = 280
X = 280/8 = 35
Number of turtles Michael bought = 60 – 35 = 25
b) Total cost of the kittens = 35 x 5 = 175

Solution to Question 191

Number of days = Jan 5 – Jan 1 = 4
Therefore by 6 A.M. on Jan 5, the clock has lost = 4 x 8 = 48 min
Till 12 noon it is 6 more hours.
Since the clock loses 8 minutes every 24 hours
In a 6 hour period, it will lose 6 x 8/24 = 2 minutes.
So the total time lost by noon on 5th January will be 48 + 2 = 50 minutes.
When the correct time is 12.00 noon, the clock will show
12:00 – 0:50 = 11:10 AM

Solution to Question 192

Let the cost of the hand bag be X
Molly had only 1/2 of the cost of the hand bag = X/2
After her mother gave her $42, she was still short of 1/4 of the cost of the hand bag.
X/2 + 42 + X/4 = X
2X + 168 + X = 4X
4X – 3X = 168
X = $168
The cost of the hand bag = $168

Solution to Question 193

Let the cot of the box of chocolates be $X
Money remaining after buying chocolates = 60 – X
With half of the money Shelly had left she buys a heart necklace.
Cost of necklace = 1/2 x (60 – X)
Remaining money = (60 – X) – (60 – X)/2 = (60 – X)/2
She then spends one fourth of the remainder on a dozen roses
= 1/4 x (60 – X)/2.
She is left with ¾ x (60 – x)/2
When Shelly returns home, she has 15 dollars left.
3/4 x (60 – X)/2 = 15
60 – X = 15 x 4/3 x 2
X = 60 – 40
X = $20
The box of chocolates cot $20

Solution to Question 194

Let T be the cost of a T-Shirt and S be the cost of a Skirt.
Then
2T + 3S = 234
And
3T + S = 120
S = 120 – 3T
Put this value of S in the equation above to get:
2T + 3 x (120 – 3T) = 234
2T + 360 – 9T = 234
360 – 7T = 234
7T = 360 – 234
T = 126/7 = $ 18
The cost of each T- shirt = $18

Solution to Question 195

Each lipstick cost $4.25 after the discount.
So the original price of the lipstick = 4.25/85 x 100 = $5
Amount of money saved on each lipstick = $5 – $4.25 = $0.75.
She bought 9 lipsticks, so the amount of money she saved = 9 x 0.75
= $6.75

Solution to Question 196

In the first two hours, the owner will have to pay $0.80
Each hour after that costs $0.15
So the amount of money paid for every extra hour can be put in a table as:

Number of hours	Cost
2	$0.80
3	$0.80 + $0.15 = $0.95
4	$0.95 + $0.15 = $1.1
5	$1.1 + $0.15 = $1.25

So the car stayed in the parking lot for 5 hours

292

Solution to Question 197

Let the cost of the pencil be X

2 pens cost as much as 15 pencils

So the cost of each pen = 15X/2

20 pens and 30 pencils cost $90.

20 x 15/2X + 30X = 90

150X + 30X = 90

180X = 90

X = 90/180 = $0.50

The cost of each pencil $ 0.50

Solution to Question 198

a) Let if X be the amount of money William has, then his sister has = 520 – X

He spent ½ of his money so he has ½ of his money left = X/2

His sister spent 1/5 of her money so she had 1 – 1/5 = 4/5 of her money left.

= 4/5 x (520 – x)

They both had the same amount of money left

X/2 = 4/5 x (520 – x)

5X = 8 x (520 – X) = 8 x 520 – 8X

5X + 8X = 8 x 520

13X = 8 x 520

X = 520 x 8 / 13 = X320

The amount of money William had was $320.

Amount of money William spent = 320/2 = $160

b) His sister had $520 – $320 = $200

His sister spent 1/5 of her money and was left with 4/5 of $200

= 4/5 x 200 = $160

His sister was left with $160

Solution to Question 199

Let my current age be X.

My Mother was 27 when I was born so she is 27 years older than me.

My Mother's current age = X + 27

8 years ago, my Mother's age = X + 27 − 8 = X + 19

This is twice as old as my age after 5 years = X + 5

So

X + 19 = 2 x (X + 5)

X + 19 = 2X + 10

2X − X = 19 − 10

X = years.

I am 9 years old now

Solution to Question 200

A book costs $16 more than a magazine.

Let the cost of magazine be X

The cost of a book = X + 16

The total cost of 3 books and 2 magazines is $68.

3 x (X + 16) + 2X = 68

3X + 2X + 48 = 68

5X = 68 − 48

X = 20/5

X = $4

Cost of the Magazine = $4; Cost of the book = 4 + 16 = $20

The total cost of 5 book and 5 magazines   = 5 x 4 + 5 x 20

                                              = 20 + 100 = $120

294

20416398R00164

Printed in Great Britain
by Amazon